William Henry Parr Greswell

Geography of the Dominion of Canada and Newfoundland

Under the Auspices of The Royal Colonial Institute

William Henry Parr Greswell

Geography of the Dominion of Canada and Newfoundland
Under the Auspices of The Royal Colonial Institute

ISBN/EAN: 9783337069841

Printed in Europe, USA, Canada, Australia, Japan

Cover: Foto ©berggeist007 / pixelio.de

More available books at **www.hansebooks.com**

GEOGRAPHY

OF

THE DOMINION OF CANADA

AND

NEWFOUNDLAND

BY

THE REV. WILLIAM PARR GRESWELL

M.A. (OXON.), F.R.C.I.

Late Scholar of Brasenose College
Author of 'A History of the Dominion of Canada'

WITH TEN MAPS

UNDER THE AUSPICES OF

THE ROYAL COLONIAL INSTITUTE

Oxford

AT THE CLARENDON PRESS

1891

PREFACE.

THE following geographical study of the Dominion of Canada and of the Island of Newfoundland is intended to be supplementary to the short history of those countries recently published by the Clarendon Press. It has been compiled from the latest and most trustworthy maps and statistics, and has been corrected throughout by the same gentlemen of the Educational Committee of the Royal Colonial Institute, who have revised the history, viz. Lieutenant-General R. W. Lowry, C.B., Dr. Rae, F.R.S., and Mr. Peter Redpath. Under the ever-changing conditions of the Dominion, especially in the West, it has been a somewhat difficult task to obtain precise accuracy of details as to the increase and distribution of population, trade, and native industries. Even the boundaries of the older Provinces of Quebec and Ontario, which we might have regarded as long since fixed and permanent, have recently (May, 1890) been undergoing revision on the North. The Province of Quebec, once called Lower Canada, is no longer contained in the Valley of the St. Lawrence, but, according to the Honourable Honoré Mercier's statement on pp. 7–8 of 'A General Sketch of the Province of Quebec,' must be extended northwards across the height of lands to James' Bay, and to the watersheds of the Hudson's Bay Rivers. Further, the Province of Ontario, once called Upper Canada, has long since ceased to mean only that fertile peninsula and centre of British colonisation surrounded by Lakes Ontario, Erie, and Huron, but has been expanding northwards to the shores of Hudson's Bay, and westwards along the shores of Lake Superior to the meridian of the Lake of the Woods, and to the boundaries of Manitoba.

The geographical expression 'The North-West Territories' has also been undergoing recently some modification. In Keith Johnston's 'School, Physical, and Descriptive Geography,' 1884, the term is said to mean (p. 340) 'all the vast region of North America through which the trading stations of the Hudson's Bay Company are scattered. It extends from the boundary of the United States away north to the Arctic Ocean, and from the Inner Watershed of Labrador westward to the heights of the Rocky Mountains.' It might have been added that before British Columbia was created a Province it was regarded as part of the North-West Territories.

Now, however, the term has been narrowed down to the extreme northern parts of the continent north of British Columbia and west of longitude 100° W., and of the boundaries of Manitoba and the Keewatin District. It includes at present the judicial districts of Assiniboia, Alberta, Saskatchewan, and Athabasca, the official and governing centre of these vast domains being at Regina. The Keewatin District, over which the Lieutenant-Governor of Manitoba exercises jurisdiction, is a long strip of territory lying directly north of Manitoba, with an eastern boundary line extending from the north-east corner of Manitoba, at the meridian of Lake Winnipeg, to Fort Churchill on Hudson's Bay, and with a western boundary line following longitude 100° W. northwards.

The North-East Territories is a recent geographical term for the comparatively unoccupied regions lying to the north of the Provinces of Quebec and Ontario, and bounded on the east by Labrador, the dependency of Newfoundland. Unlike the North-West Territories, the term 'North-East Territories' has no political significance. Meantime it may be noticed that, in a Bill before the Parliament of the Dominion of Canada to amend the 'North-West Territories' Act, a clause has been inserted, at the instigation of the Government, changing the name 'North-West Territories' to 'The Western Territories of Canada,' the reason given for this change being that the use of the word 'North' was a misnomer, and conveyed a wrong impression to intending settlers. Ultimately, no doubt, these Territories, as they are claimed by the nearest Provinces, or separately occupied and inhabited, will lose their customary and historical nomenclature. An unorganised and unoccupied territory quickly becomes a judicial district, and in due course of time a Province. In the future a series of fair and fertile Provinces will arise upon the Prairies of the North-West Territories, fully organised, surveyed, and equipped with local and provincial autonomy. At present, however, and perhaps for some time to come, the geographical nomenclature adopted in the text may stand.

The orthography of certain names and places still seems a matter of doubt and uncertainty, a fact which may be explained by the presence of a French-speaking community alongside of British colonists. Writers on

Canadian history have not yet adopted in every case a uniform spelling. I have spelt Mines thus instead of Minas, in accordance with the fashion set by Mr. Kingsford, the latest historian of Canada, although the precedent of the poet Longfellow is against this. On p. 37 I have suggested that the County of Richelieu, like those of Charleroix, Montcalm, and Joliette in the Province of Quebec, was so called to commemorate a great Frenchman. Hayden and Selwyn have remarked in their 'Geography of North America,' p. 459, that it was so named, like the County of Terrebonne I presume, to denote simply the richness of the place. Canadian Rivers also occasionally have two names, as the Dauphin or Little Saskatchewan, the St. Charles or Assiniboine, the Souris or St. Pierre, the Peace or Unjigah, a fact which may sometimes puzzle the student. Moreover, Canada would seem to be as famous for its patron saints as Cornwall in our own country. St. John, for example, is commemorated so often in the New World that the reader is apt to be puzzled.

Although many remarks have been made in the text on the ocean currents, winds, rainfall, atmosphere, climate, soils, and general physical conditions of the country, there is no attempt made to arrange these conditions under any general system. The first chapter is simply a very brief and prefatory account of some of the more striking features of the land. In the last chapter a few remarks have been offered on the industries, wealth, and social progress of the country, taken chiefly from 'Agricultural Canada: a Record of Progress,' by Professor W. Fream, LL.D., and published under the direction of the Government of Canada (Department of

Agriculture), 1889; also, from an 'Official Handbook of Information relating to the Dominion of Canada,' published by the Government of Canada (Department of Agriculture), 1890; and from 'British Columbia, its Resources and Capabilities,' reprinted from 'Canada: a Memorial Volume,' Montreal, 1889. For the paragraphs on educational progress I am indebted to a paper read by Mr. Henry F. Moore before the Fellows of the Royal Colonial Institute, April, 1889, entitled 'Canadian Lands and their Development.' For general information on the Dominion I have consulted the following papers read before the Fellows of the Royal Colonial Institute:— 'The British Association in Canada,' read by Sir Henry Lefroy, K.C.M.G., C.B., January, 1885; 'Newfoundland: our oldest Colony,' read by Sir Robert Pinsent, D.C.L., April, 1885; 'Recent and Prospective Development of Canada,' read by Mr. Joseph G. Colmer, C.M.G., January, 1886; 'British Columbia,' read by the Right Rev. the Bishop of New Westminster, March, 1887.

My object throughout has been to obtain the most trustworthy information from official sources, and from the accounts of those who have had the most recent personal experience of the Provinces of the Dominion. Occasionally there are disputed points in the geography as there are in the history of the country. The climate itself is often variously described by those who have lived in the Dominion. This is only natural when we consider the vast area and extent of the country, exceeding, if we take into account the water surface of the lakes, the whole area of Europe. To convey correct ideas on this point, a few remarks have been offered in the text on the climate of each Province. The climate of Hudson's Bay has often been a *vexata quaestio* with experienced sailors. Some years, no doubt, the waters of this bay are navigable for a longer time than others; and it is impossible to foretell with exactness what kind of winter will prevail in the Arctic regions, or the number of days in the year when any particular waters will be open. The nature and extent of ice in Hudson's Strait cannot always be known beforehand, but as the Hudson's Bay route to the interior of the continent is more generally used, the climatology of these regions will be better known. On another point, viz. the northern limit of the grains and grasses in the Dominion, various opinions have prevailed, but perhaps it is safer to put the wheat limit at 58° N. As in other countries so in the North-West Territories of Canada, there are, even at the same levels, both sheltered and exposed places. In certain chosen localities, favoured by soil and sun, wheat and barley may ripen at high latitudes, even beyond 61° N. latitude, but it is not safe to argue that at every spot along the same parallels of latitude equally favourable results will be obtained. The summer isotherms, however, of the Dominion are deserving of very careful study before we pass judgment on the climate of the Dominion, even under the Arctic Circle. Further, with regard to the wheat-bearing zone, it must be remembered that some kinds of wheat, notably the Onega and Saxonka varieties, will ripen more quickly and more surely than the usual kinds. Generally speaking, it must be assumed that there is no such ζείδωρος ἄρουρα, to use an Homeric phrase, as this in the world. Nowhere is healthy colonisation and settlement carried on so quickly. In my 'History of the Dominion'

I have pointed out, on p. 263, in a table quoted from 'The British Association in Canada' (Proceedings of the Royal Colonial Institute, vol. xvi. 1884-5), that by far the largest portion of the immigrants have been recruited from the British Isles, and latterly from Iceland and Scandinavia. In 1888 the French and Belgian immigrants only numbered 255 out of a total of more than 28,000, the Germans only numbering 403. The filling-up process is going on, of course more especially in Manitoba and British Columbia and the adjoining judicial districts.

The extraordinary traffic which passes down the Sault Ste. Marie Canal, and promises to exceed greatly that of the Suez Canal, is a remarkable proof of the growth and wealth of the Far West. This canal passes through United States Territory, but, as pointed out on p. 48, the Canadians are making a canal of their own. This will be another and extremely important link in that system of unrivalled waterways and canals which exists in the Dominion. On p. 321 of the 'History of the Dominion' I have given a list of the canals of the Dominion with their mileage, gathered from 'Canada: a Statistical Abstract and Record,' 1887. In addition to authorities already cited I have quoted, for the purpose of general and descriptive accounts, from Sir W. Butler's 'Wild North Land' and 'The Great Lone Land;' from 'Picturesque Canada,' edited by the Very Reverend G. M. Grant, Principal of Queen's University, Kingston, Ontario; from 'Hochelaga,' by Sir F. Head; from 'Manitoba, its Growth and Present Condition,' by Professor Bryce; and from 'Canadian Pictures,' by the Marquis of Lorne. With regard to the latter authority I have ventured to alter, in a quotation on p. 48, the words 'if the ships be of 1400 tons' to 'a ship of small tonnage,' some of the canals having only nine and not fourteen feet of water over the sills of the locks. I have also quoted occasionally from Lord Dufferin's speeches, and from the pages of Parkman, the well-known historian of Canada. For kind supervision and invaluable help I am indebted to my revisers and to Mr. J. S. O'Halloran, the Secretary of the Royal Colonial Institute. For special information on the subject of Newfoundland I have to express my thanks to Sir Robert Pinsent, and for the latest statistical and other information on the Provinces of the Dominion, I am largely indebted to the gentlemen of the High Commissioner's Office in London. With regard to maps and geographical nomenclature, it will be seen that I have not found space on the maps for every place mentioned in the text, but a sufficient number, perhaps, have been given for the purposes of guidance and illustration.

WILLIAM GRESWELL.

DODINGTON,
June 17, 1890.

CONTENTS.

CHAPTER I.
	PAGES
The Geography of the Canadian Dominion	1–28

CHAPTER II.
The Province of Quebec	28–41

CHAPTER III.
The Province of Ontario	42–51

CHAPTER IV.
The Province of Nova Scotia	51–58

CHAPTER V.
The Provinces of New Brunswick and Prince Edward Island	58–66

CHAPTER VI.
The Province of Manitoba	67–73

CHAPTER VII.
The North-West Territories	73–81

CHAPTER VIII.
The Province of British Columbia	81–93

CHAPTER IX.
The Island of Newfoundland	93–102

CHAPTER X.
Industries, Wealth, and Social Progress	102–123
Appendices	126–141
General Index	143–154

LIST OF MAPS.

			PAGE
I.	General Map, showing the boundaries of the Provinces, with Railways	To face	1
II.	Historical Map, showing Ancient Settlements	,,	7
III.	Physical Map, showing cultivable zone of grains and grasses, sage country, region of droughts, summer and winter isotherms, cold and warm ocean currents, the prairie steppes	,,	17
IV.	The Province of Quebec .	,,	29
V.	The Province of Ontario	,,	43
VI.	The Provinces of Nova Scotia, New Brunswick, and Prince Edward Island	,,	51
VII.	The Province of Manitoba, with southern portion of the North-West Territories .	,,	67
VIII.	The Province of British Columbia .	,,	81
IX.	The Island of Newfoundland . .	,,	93
X.	The Townships of the North-West .	,,	112

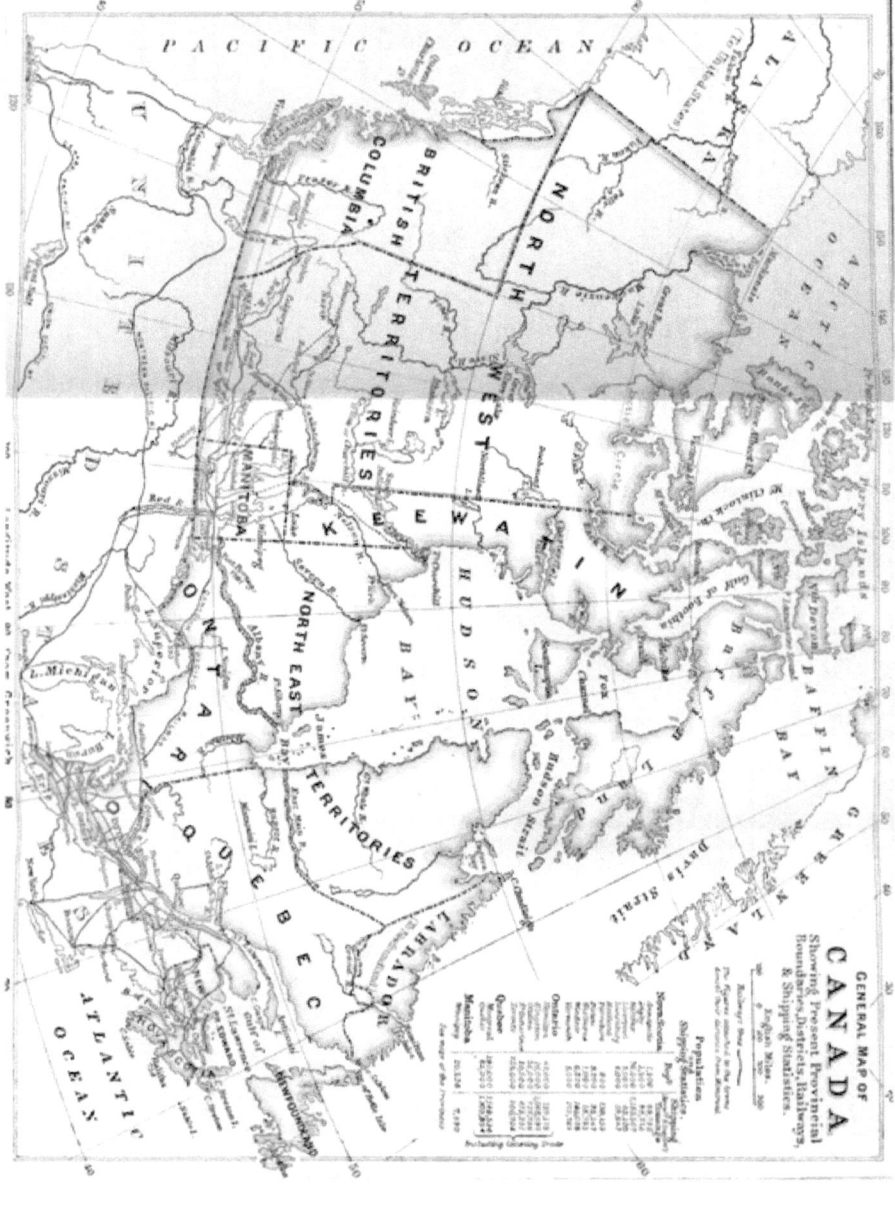

CHAPTER I.

The Geography of the Canadian Dominion.

(1) AFTER its first discovery by Cabot (1497) the continent of North America remained for many years untenanted by genuine colonists and settlers. Time alone could prove its complete adaptability to the nations of Northern Europe, and convince French, English, Germans, and Scandinavians, that here along the same lines of latitude was the natural home for the development of their energies. In the first place, the very nature of the climate and its sea-coast was suited especially to the life of a seafaring folk. The great island of Newfoundland, the bulwark of the St. Lawrence Gulf, is indented with numerous bays and inlets, and boasts of a larger coast-line for its size than any other important island in the world. Nova Scotia, New Brunswick, and Prince Edward Island have peculiar attractions for fishermen and adventurers. To explore them thoroughly from bay to bay, from fiord to fiord, in the ardour of first discoverers, must have been a fascinating occupation. And these maritime provinces lay at the threshold of undiscovered worlds beyond. The St. Lawrence was a grand natural highway, which took the traveller into the very heart of a country abounding in natural wealth. There was nothing in the known world to equal the marvellous chain of the great lakes. To those who had been accustomed to look with wonder and surprise upon the small inland lakes of Great Britain, nestling amongst picturesque hills, how noble these great inland seas, fringed with forest wealth to their edges, must have

appeared! As lake beyond lake was discovered, and this splendid natural highway was opened up, the heart of the colonist and trader rejoiced. In dry countries like South Africa and Australia the main want is water and means of transport : and places only a few miles distant from one another are effectually cut off and isolated by a tract of country where it is difficult, nay, impossible, to construct a road. Before the days of railways, the colonists in South Africa and Australia were debarred from all the innumerable benefits which come from easy and quick communication. The merchant needs speedy distribution for his produce, the farmer an easy road to his market, and the politician, who guides the affairs of the colony, needs to be as quickly as possible in touch with all sections of the community.

(2) In Canada the great lakes give all classes the means of communicating with one another, and out of this natural advantage has arisen the extraordinary prosperity of the Dominion. To gain an idea of the vastness of the Lake Districts of Canada, we need simply refer to the map and see how wide is the area they occupy. Lake Superior is the largest sheet of fresh water in the world, being 420 miles from east to west, with an average breadth of 80 miles. The St. Lawrence and the Canadian lakes are estimated to hold 12,000 cubic miles of water, or more than half the fresh water on the globe. The shores of the Dominion are not mere barren beaches such as border the sea, but hundreds of miles of green fir-clad banks along which vast and solemn pine-forests grow, and the lumberer's axe finds ready and profitable spoil. In spite of the rigour of the climate, the country proved to be extremely healthy for Europeans, and here more than elsewhere in the New World the stock has thriven. For races dwelling in the north of Europe, Canada is a natural habitat, especially for the Bretons

and Scots. Steam has come as a powerful auxiliary, and has made the most of the advantages already at hand. Without losing any of its native fibre and hardihood, as it must do in tropical colonies, the British stock has lived and prospered, and heaped up wealth. There is a greater equality of wealth amongst all classes, and a more evenly distributed population in the Dominion, than in Australasia or South Africa. The lakes and rivers are partly the cause of this. They have encouraged dispersion by the facilities of travel and transport. Especially during summer their open waters have provided a cheap and easy transit. In the colonies of the South Pacific we hear of congested towns and crowded thoroughfares, and all the ills that, along with many advantages, centralisation ever brings in its train. Many railways exist there, and their mileage is constantly being increased; still they are narrow and expensive arteries, compared generally with the placid lake and broad river as a medium. There is, therefore, no such town as Melbourne or Sydney in the Dominion.

(3) The industries of a new country also determine the character of the population. The great mining centres of Australia and South Africa localise trade, and cause a divergence of population towards themselves. In Canada the occupations of agriculture, of lumbering and of fishing, which are the chief ones, spread the people abroad over many country towns and villages. The purely rural population of Canada is extremely numerous, and constitutes a most healthy and substantial class. It is the 'mascula proles docta ligonibus' who so often have proved the mainstay of a country. The seamen and fishermen of Newfoundland, New Brunswick, and Nova Scotia are the hardiest perhaps in the world, and reproduce the stern and wholesome qualities of Scandinavian ancestors.

(4) As useful waterways in connection with the St. Lawrence we must consider the Ottawa, St. Maurice, Saguenay on the left, and Richelieu, St. Francis, Chaudière on the right bank. To the north lie the vast inland waters of Hudson's Bay stretching 600 miles from east to west, and 1300 miles from north to south, and providing a waterway for five months of the year to the very heart of the continent. Port Nelson, the trade entrepot on the west shores, is only 2941 miles distant from Liverpool, being 100 miles nearer than New York[1]. Thence by river, lake, and portage, for hundreds of miles the way lay open to the utmost bounds of the Great West. For 200 years fur-traders have explored the continent and planted their Forts. From Hudson's Bay Franklin undertook his land journey to the Polar Seas, and from Hudson's Bay the Selkirk colonists started to found the Prairie Province of Manitoba.

(5) From the mouth of the St. Lawrence, and from Hudson's Bay, explorers could press westward to solve the geographical problems of a new continent, till beyond the sources of this great river and the latitude of Lake Superior, they found the ground rise steppe above steppe[2], and so reached the huge barrier of 'the Rocky Mountains.' This mountain chain runs down the whole continent for 3000 miles from north to south. Once this range and its subsidiary elevations were passed, the European found a more genial land and milder hillsides abutting on the great Pacific, and facing the eastern world with its countless nations and peoples. Here, too, was found an island—Vancouver—guarding the entrance to this northern land on one coast, just as Newfoundland guards it sentinel-like at the entrance to the St. Lawrence on the other. On these two islands floats the British

---

[1] See p. 317 of Hayden and Selwyn's 'North America.'
[2] Appendix I.

flag, and at either end, at Halifax and Esquimault, is stationed a squadron of British men-o'-war, keeping watch over the ocean paths, and over that vast limb of England's empire, which has grown so mighty since the day when the sailors first crept cautiously into the St. Lawrence, hoping, perchance, to find that North-West Passage which had filled the imagination of Columbus when he dreamt of Cipango and Prester John. Here in truth is the true North-West Passage, not a permanent sea-way, but a quick communication by ship and rail.

(6) The term Canada is said to have been an Indian word, Kamatha, meaning a collection of huts, which the French discoverers applied to the country around the St. Lawrence. At the present time it is used somewhat loosely to express the whole of the Canadian Dominion, but Canada Proper was originally the geographical expression for the watershed of the St. Lawrence from its mouth to the Lake of the Woods, a distance, roughly speaking, of 1300 miles from east to west. It was not so used by Cartier in 1534, being limited more precisely by him to the central part of the territory between Montreal and Anticosti[1]. In the French annals of colonisation it may be regarded as the northern part of New France, distinguishable on the east from the Island of Newfoundland and Nouvelle Bretagne or Labrador and the regions included under the term Acadia, on the south from French Florida and Louisiana. According to Sir W. Logan's report, dated May 1863, Canada Proper, meaning the basin of the St. Lawrence, has an area of 530,000 square miles, being nearly five times as large as Great Britain.

(7) In the early history of North America charters and grants given by both French and English kings to companies and individuals were constantly altering and

[1] Kingsford, vol. i. p. 3.

upsetting previous boundary lines and landmarks. In the first instance there was a grand simplicity about Pope Alexander's bull (May 1493), the year succeeding Columbus' discovery of the Bahamas, in which he drew an imaginary line from the North to the South Pole, a hundred leagues west of the Azores, assigning to Spain all that lay to the west of that boundary, while all to the east of it was confined to Portugal. This was virtually handing over to Spain by papal sanction the whole of North and South America, excepting that small part of the latter which lies, roughly speaking, to the east of the mouths of the Amazon, and is contained in the Republic of Brazil. This imaginary geographical definition was not accepted by other nations. Francis I of France exclaimed, 'What! shall the kings of Spain and Portugal divide all America between them, without suffering me to take a share as their brother? I would fain see the article in Adam's will that bequeaths this vast inheritance to them.' So in 1524 Verrazano, commissioned by the king, conducted a series of explorations along the sea-board from Cape Fear to the Bay of New York and Narragansetts, and called the country *New France*. In 1534 Cartier set up on the Peninsula of Gaspé both the cross and the arms of his country claiming the valley of the St. Lawrence, a river so named by him on August 10, 1535, the martyr's anniversary.

(8) Further north it must be remembered that Cabot, by his discovery of Labrador and Bacalaos (Newfoundland, 1497), had given England a prior claim to the continent of North America by right of discovery, although it was not till 1583 that Sir Humphrey Gilbert unfolded the flag of England on the Island of Newfoundland and exacted homage from the fishing fleet. The Spaniards and Portuguese, it will be seen, left this

HISTORICAL MAP SHOWING ANCIENT SETTLEMENTS.

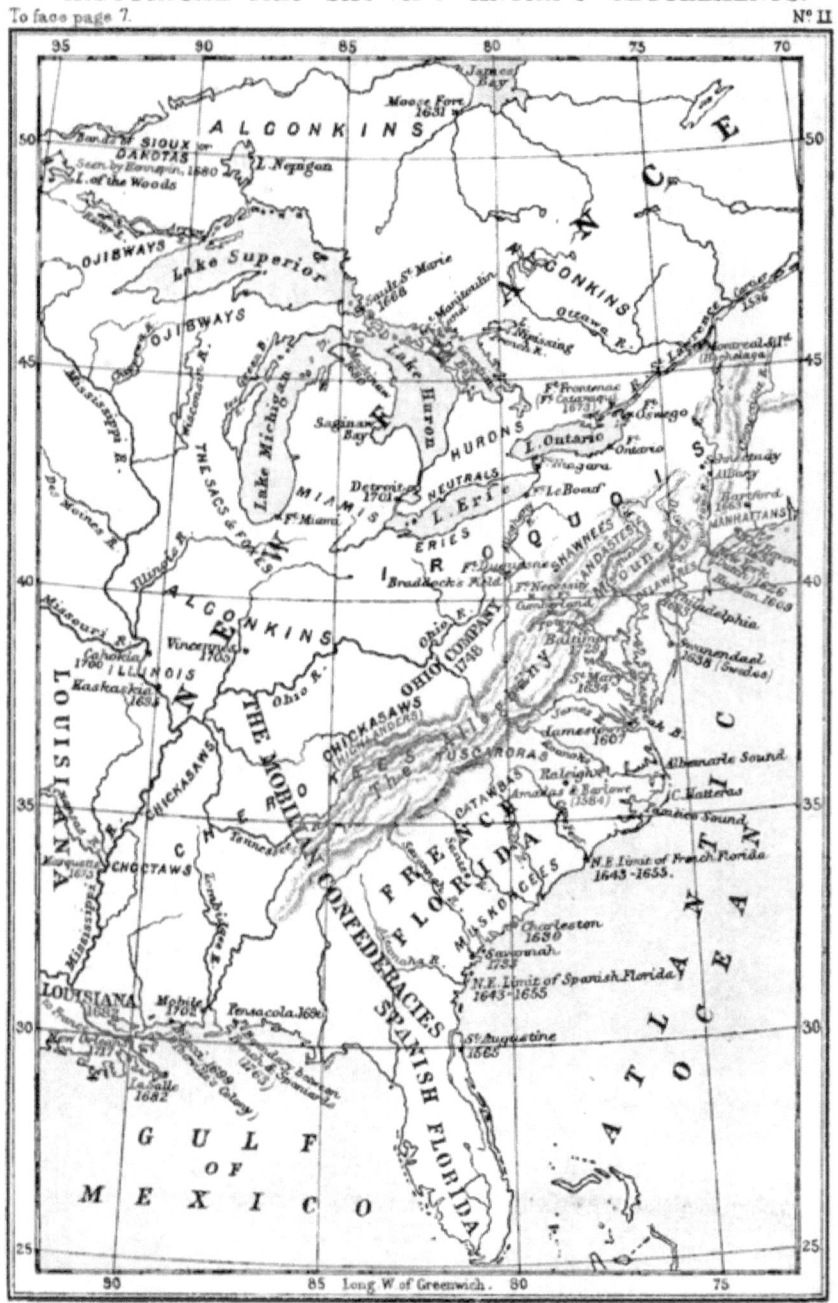

part of North America almost entirely to France and England. Cortereal, the Portuguese sailor, reached in 1501 the Gulf of the St. Lawrence and explored the North American coasts up to 50° N. He is said to have freighted his vessels with more than fifty Indians whom he sold as slaves. As the historian Bancroft remarks, Labrador (from Laboratores or slaves transferred from the territory south of the St. Lawrence to a more northern coast) is a memorial of his voyage and the only permanent trace of Portuguese adventure within the limits of the North American continent, just as Portugal Cove is said to have been the only sign of their fishing expeditions in the neighbouring island of Newfoundland [1].

(9) In 1540 Francis de la Roque, Lord of Roberval, obtained from the French king a commission as Lord of the unknown Norumbega (comprising the littoral and mainland of New Brunswick) and Viceroy, with full regal authority over the immense territories and islands which lie near the Gulf and along the river of St. Lawrence. This clashed with the Cabots' Charters, given them by Henry VII and Henry VIII, and assigning them all lands that might be discovered with sole right to trade. In June 1578 a Royal Charter was given to Gilbert with 'licence to discover... such remote and barbarous lands not actually possessed of any Christian Prince or People, as to him, his heirs and assigns... shall seem good, and the same to have, hold and occupy and enjoy to him, his heirs and assigns for ever, with all commodities, jurisdictions and royalties both by sea and land;' also, according to the Charter of March 1584, Walter Raleigh obtained even greater concessions by sea and land. He had a monopoly of all commodities, jurisdictions, royalties and privileges by sea and land, and power to repel all intruders

---

[1] According to some authorities Labrador is so named after La Bradore, a Spanish captain.

that came within 200 leagues of his settlement or dwellings. In July 1584 his captains, Amadas and Barlowe, took possession of Roanoke, an island of fifteen or sixteen miles in length, and a beginning of colonisation was made. Virginia meant all the country between the French dominion of Canada and the Spanish dominion of Florida. Here was the beginning of the power that wedged itself firmly between the northern and southern parts of New France and held the seaports and littoral of the east. The Virginia Patent discarded Verrazano's barren proclamation along the coast, but did not touch Cartier's acquisitions to the north.

(10) In 1606 the Virginian settlements which had failed at first were revived. They received the right, according to King James' Patent, to have and hold all the land from Cape Fear to the St. Croix River. This territory was divided into two districts: (1) the north part, controlled by the Plymouth Company, and taking in all land from east to west between latitude 41° N. and 45° N., thus including the coast-line from the mouth of the St. Croix to the Hudson River, and embracing the Lake country in the interior; (2) the London Company included the littoral from the mouth of Cape Fear River to the mouth of the Potomac, lying between latitude 34° N. to 38° N. Jamestown was founded by the London Company in 1607, and the two capes at the entrance of the bay were named after Henry and Charles, King James' two sons. This was the first *permanent* settlement of Englishmen in the continent of North America. It may be remembered that the first permanent French colony had been settled at Port Royal (Annapolis) by De Monts in 1604, three years previously. In 1620 the Plymouth Company was reorganised as the Council of Plymouth for New England, and included within their boundaries the whole North American coast

from latitude 40° N. to 48° N., thus embracing the whole of Canada from Tadousac along the valley of the St. Lawrence to the interior.

(11) It must be borne in mind that in 1598, more than fifty years after Roberval and Cartier, a charter was given to De la Roche by the French king. He was Viceroy of Canada, Acadia (meaning Nova Scotia, New Brunswick and part of Maine) and the adjoining territory, with sole right to carry on the fur-trade. In 1603 an exclusive Patent was given to De Monts from 40° N. to 46° N. latitude. To this territory and more beyond it the Plymouth Company laid claim in 1620. Thus one charter is constantly overlapping and including another, and the confusion of rights could only be settled by the strong arm. It may be however some help to notice here what New France meant at different dates, before the name disappeared altogether from the North American continent.

(12) In 1655 New France meant, according to French claims, Newfoundland, Labrador, Acadia and Canada, and the country beyond as far as it was discovered to the west and south by the Jesuits and others. It also meant French Florida[1]. This was a tract of country wedged in between Virginia on the north and Spanish Florida on the south. The Cape Fear River on the north and the Altamaha River on the south formed definite geographical limits. It was a tract of country including the watershed of the Savannah, Santee, and Great Pedee Rivers. Between the St. Croix River and Cape Fear River, that is between Acadia and French Florida, lay New England, from St. Croix to the Connecticut and Hudson, New Netherlands along the Hudson and Delaware, Maryland and Virginia from the Susquehanna to Cape Fear River. Along the interior the

[1] See Labberton's 'Historical Geography.'

Alleghanies formed a boundary between New France and the Coast Settlements.

(13) In 1713, by the terms of the Treaty of Utrecht, New France lost the Atlantic sea-board she possessed in Newfoundland, Hudson's Bay, and also Nova Scotia and New Brunswick; but she still retained the valley of the St. Lawrence, Cape Breton, St. John's Island, (Prince Edward Island), and the vast region lying on either side of the Mississippi from the Lakes to the Gulf of Mexico, known as Louisiana, claimed for Louis XIV by De la Salle (1682)[1].

Louisiana was an empire in itself, bounded by Oregon territory on the north-west and New Spain on the west. After February 10, 1763, New France disappeared entirely. The settlements comprised in the St. Lawrence valley, the Lake region, the Great North-West with its vast hunting preserves, and the command of the whole country southwards as far as the Gulf of Mexico, with the Mississippi as the western boundary, had fallen into English hands. This was the land Voltaire said was 'not worth fighting for, because, after all, it only consisted of a few acres of snow.'

(14) After September 3, 1783, the United States Territories appear upon the map, and the dividing international line is shown from St. John's River and the coast to longitude 95° W. or about the meridian of the Lake of the Woods. In 1803, Napoleon sold Louisiana to the American Government for £3,000,000, more than doubling the area of the Republic. The international boundary line then appears (lat. 49°) upon the map, twenty degrees further west, as far as long. 115° W. up to the Oregon territory. This boundary line was completed to the Pacific coast after the acquisition of Oregon territory from Great Britain in 1846.

[1] Appendix II.

There were two minor disputes, already alluded to, between Canada and the United States on the question of boundaries; (1) that touching an area of 12,000 square miles between the Province of New Brunswick and the State of Maine, settled in 1841 by the Ashburton Treaty; (2) that touching the possession of the island of San Juan, decided in favour of the United States. Alaska was bought from Russia by the United States in 1868. It was not till 1873 that the present boundaries of the Canadian Dominion appeared upon the map.

(15) The Southern boundary line, as it now exists between British America and the United States, follows the $49°$ north latitude from British Columbia to the Lake of the Woods, thence in a wavy line along Rainy River to Thunder Bay on Lake Superior. Thence to the east through Lakes Superior, Huron, Erie, Ontario till the $45°$ line is touched upon the St. Lawrence. The State of Maine is thrust northwards, forming a rough-shaped triangle, with New Brunswick on its eastern side. The $45°$ line is again touched at the Bay of Fundy. At the extreme north-west, British North America is separated from Alaska along the $141°$ of west longitude. The Dominion of Canada comprises, therefore, the whole of the continent north of the United States frontier, with the exception of Alaska, stretching up to the Arctic regions, bounded on the east by the Atlantic and on the west by the Pacific oceans. It includes the following Provinces and Territories:—Quebec, Ontario (Canada Proper), Nova Scotia, New Brunswick, Prince Edward Island, Manitoba, North-West Territory, British Columbia.

(16) The **Area** of the Dominion, not including the Lakes, is 3,470,253 square miles, the greater part being in the North-West Territories. This is nearly equal to that of the United States, which with Alaska is 3,585,900 square miles, while that of Europe is computed at 3,800,000.

The following list shows the estimated population of the most important cities in the Dominion (1890)[1]:—

Montreal	. . .	202,000
Toronto	. . .	172,000
Quebec	. . .	65,000
Ottawa (capital)	. .	44,000
St. John (N. B., with Portland)	. .	44,000
Hamilton	. . .	43,000
Halifax (N. S.)	. .	40,000
London	. . .	27,000
Winnipeg	. . .	22,000
Kingston	. . .	17,000
Victoria (B. C.)	. .	14,000
Charlotte Town	. .	13,000
Brantford	. . .	13,000

Compared with Australia the urban, as distinguished from the rural, population is far less in proportion. For instance, the city of Melbourne, the capital of the colony of Victoria, is now said to contain a population of 410,000 out of a total of 1,000,000; and the city of Sydney, the capital of New South Wales, contains 351,000 out of a total of about a similar number. Mr. Macfie, an Australian colonist, has recently pointed out that this crowding of one-third of the populations of New South Wales and Victoria in Sydney and Melbourne respectively is an unhealthy sign. 'How striking is that abnormal and unproductive concentration of an excessive proportion of the inhabitants of Australia in a few towns, compared with the wholesome distribution of population in the most prosperous countries of Europe and America where land culture is properly held to be the chief industry. In the United States in which agriculture and horticulture take their rightful place, less than one-seventh of the total population is diffused over twenty-four cities, each containing inhabitants exceeding 70,000. Some idea of the importance of American agriculture may be formed from the fact that it produces an annual yield of nearly £800,000,000, and employs on 5,000,000 farms 10,000,000 persons. Sweden and Norway, with 6,000,000 industrious people, have only half a million living in towns, the

[1] See 'Official Handbook,' published by the Government of Canada, January, 1890.

remaining 5½ millions being thrifty, hard-working peasants[1].'

The population of the Dominion of Canada was distributed thus, according to the last Census :—

	Males.	Females.
Ontario	976,461	946,767
Quebec	678,109	680,918
Nova Scotia	220,538	220,034
New Brunswick	164,119	157,114
Prince Edward Island	54,729	54,162
Manitoba	37,207	28,747
The North-West Territories	28,113	28,333
British Columbia	29,503	19,956
	2,188,779	2,136,031
	4,324,810	

It is now (1890) calculated to exceed 5,000,000. We may argue, therefore, that a very large proportion of the colonists are agriculturists, fishermen, and farmers. This is a more wholesome sign of a country's wealth, a prosperous rural population being regarded at all times as the 'backbone of a country.' In the question of the distribution of the population, it must be remembered that Canada's unrivalled lakes and rivers have helped largely to create many points of industry in all quarters[2].

The growth of the town populations has been remarkable nevertheless. For instance :—

In 1801 the population of Toronto was	336
,, 1830 ,, ,,	2,860
,, 1845 ,, ,,	19,708
,, 1851 ,, ,,	30,775
,, 1881 ,, ,,	86,445

In 1816, Quebec contained 14,880 inhabitants; in 1851 it contained 42,052; in 1881, 62,446. Montreal, fifty years ago, held less than 30,000 with suburbs; it is now the largest city. Winnipeg provides us with a

[1] Extract from Paper entitled 'Aids to Australian development,' read before the Fellows of the Royal Colonial Institute, December, 1889.   [2] Appendix III.

more recent example of quick growth. In 1888 its population numbered 25,000, in 1870 it was only 200. Contrasted with Cape Town, the capital of a country which has been colonised for 300 years by Europeans and only numbers 30,000 white inhabitants, this growth in Canada, although it does not touch the rate of the Australian cities, is very remarkable. It is calculated that British North America could easily support a population of 140,000,000 at the rate of 40 to the square mile. The population of England is 450 to the square mile.

(17) **Physical Features.—River Systems.**—The Canadian Dominion is pre-eminently a land of wide lakes and long rivers. There is no such waterway as that of the St. Lawrence in the world. From Belle Isle to the head of Lake Superior, a distance of 2400 miles, there are only 72 miles of canal. Ocean-going steamers can steam 1300 miles up its course into the country. At its mouth (usually placed between Cape Chatte on the southern bank and Pointe des Monts on the north) the St. Lawrence is 20 miles across. There are four main *river systems* in the Dominion: (1) That of the St. Lawrence; (2) the Nelson River with Lake Winnipeg, into which the Saskatchewan, Assiniboine, Red, and Winnipeg Rivers find an outlet; (3) Mackenzie River—each of these draining a basin of half a million of square miles on the average; (4) the Fraser River, in British Columbia. For purposes of inland communication these systems are invaluable, but in addition the Dominion has deeply indented shores on the Atlantic and Pacific sides, and a coast-line of many thousand miles, along which are the most profitable sea-fisheries in the world. The products of the fisheries, both exported and used for home consumption, were calculated (1888) to be worth £6,000,000[1].

(18) **Plains and Mountains.**—On the east there are

[1] 'Official Handbook,' 1890.

a series of plains, divided from one another by comparatively low ranges. In the centre is the northern slope of the great central plain of North America, so flat and even that for many miles it seems possible to drive a carriage straight ahead without natural barriers. A small tumulus seems a hill on the vast unbroken level. Near Winnipeg, a small rising called 'Stony Mount,' of 60 to 80 feet high, is the only elevation along an immense river valley. The blizzards, or sudden and severe snowstorms, which are one of the drawbacks of the American continent along its interior plateaux, are much less frequent north of $49°$ of latitude. No doubt the lower the plateaux the less liable are the inhabitants to this terrible scourge. It is in the United States Territories that blizzards are chiefly felt.

On the extreme west is a rough mountainous country, including the northern portion of the Rocky Mountains and Cascade Mountains. The geographical features of British North America are simply a continuation, although under very important variations of climate and latitude, of those of the United States. As there are plains in Texas, Montana, Kansas, Dakota, Nebraska, so there are in the prairies of the Great North-West. As there is a Pacific slope in United States Territory distinguishable in features and climate from the interior plateaux, so there is a Pacific slope, only of greater width, in British Columbia. The Rocky Mountains are the Great Divide, the high wall cutting off one part of the continent from the other. Mount Brown and Mount Hooker are the highest peaks in the Dominion, and reach the altitude of 16,000 and 15,690 feet respectively.

(19) **Lakes.**—It is however its magnificent *lake system* which is the peculiar characteristic of British North America. There are three main lake systems lying upon the surface of the country almost in a straight line from

south-east to north-west. Confining our attention simply to the largest sheets of water, under Group I we may place Lakes Ontario, Erie, Huron, and Superior; under Group II, Winnipeg and Winnipegosis, and Manitoba; under Group III, Lakes Reindeer, Athabasca, Great Slave, and Great Bear. The extent of country through which this chain of lakes stretches is from latitude 40° N. to the Arctic Circle. The minor lakes and lakelets of the Dominion may, in each system, be counted literally in thousands. For the sportsman, fisherman, and naturalist, they provide, in many cases, a most picturesque and productive recreation ground. If we except, perhaps, the newly discovered lakes of Equatorial Africa, there are no lake systems anywhere in the world approaching those of North America. In the account of the Provinces the lakes will be treated more particularly. Michigan, of course, lies outside Canada.

(20) **Climate and Products.**—The Canadian Dominion extends from the latitude of Rome to that of the North Cape in Norway, and has consequently a great variety of climate. Generally speaking, it is much colder in winter and warmer in summer than in the British Isles. The average summer temperature of England is about 60–62°, that of Canada between 60° and 70°. Taking a few points from east to west, we shall find that the heat is greatest in the interior plateaux, as we might expect in a continental climate. The average summer temperature of Halifax is 60°, of Fredericton 64°, of Quebec 69°, of Montreal 70°, of Winnipeg 65°, and of Vancouver 61°. In the United States, further south, we shall find the summer hotter than at any of the above places. Central Illinois has an average summer temperature of 74°, Ohio of 70–74°, Iowa of 72–78°, Kansas and Missouri are hotter still. These latter temperatures are ten degrees too high for wheat, which ripens best at 60–65°. With a

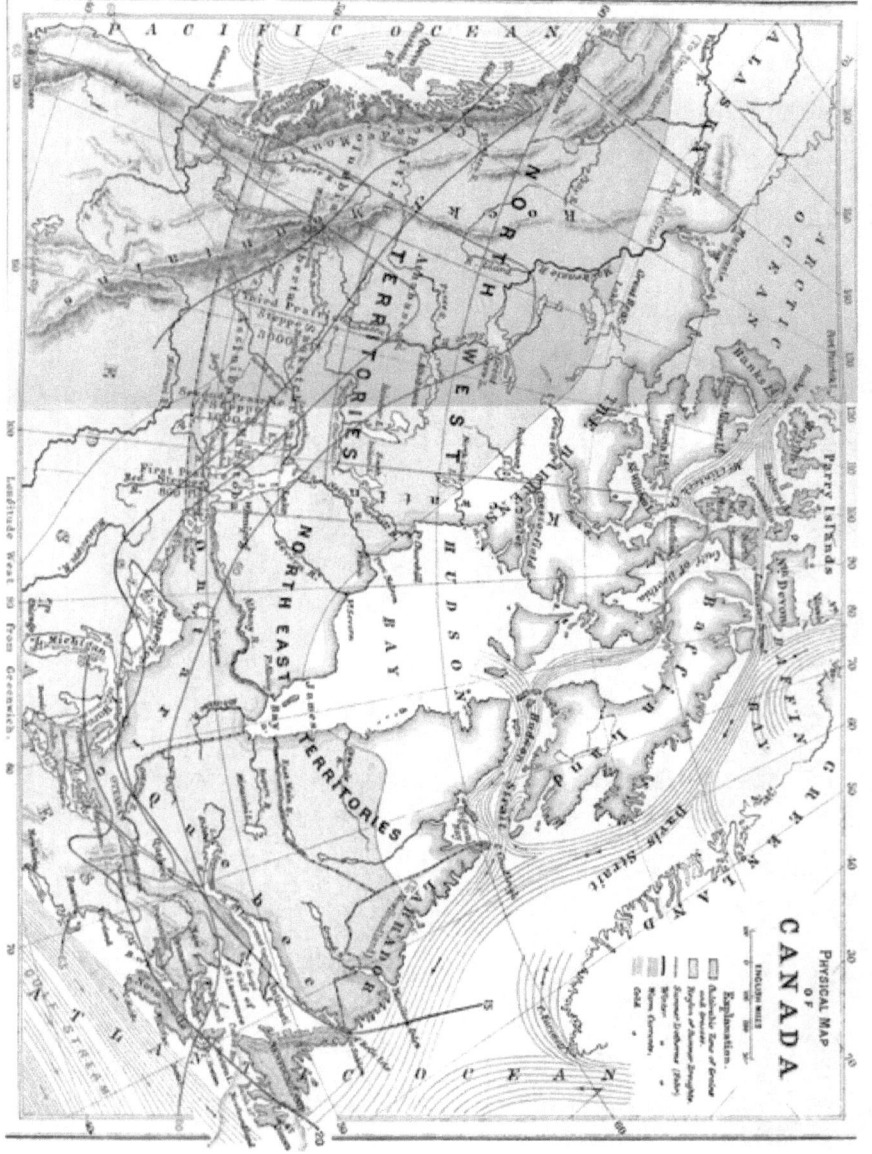

temperature above 70° it withers up, and ripens prematurely. Although situated further south, Kansas, Nebraska, Dakota, and Minnesota, in the United States, are often colder in winter than the Canadian provinces of Quebec, Ontario, and Manitoba. The apparent anomaly of a milder climate further north in winter may be explained by the fact that the interior plateaux of North America are very lofty in Mexico, and retain their elevation through the United States till they reach the Canadian frontier. They then sink to 800 and 600, and even 400, feet. It is well known that one mile in height (5280 feet) means a difference of fifteen degrees in temperature. The variation of temperature within a space of twenty-four hours is great in Canada, but it is more remarkable in the desert areas of the United States. There the thermometer will drop from 80° or 90° during the day to freezing point at night. This is prejudicial both to plants and animals. The country to the west of the 98th meridian in the United States Territory is chiefly a desert, owing to the absence of summer rains. The soils also are deeply impregnated with salts and alkalies destructive of vegetation. The great central prairies of the Canadian Dominion have, therefore, many climatic advantages over the United States to the south. The temperature is more even, the soil is richer, the alkalies disappearing as you travel north across the border, and the rainfall in summer is more plentiful. Here is the best wheat area in the world, extending over 1,400,000 square miles of country. Grasses and the coarser grains can be raised over more than 2,000,000 square miles. In the warmer parts maize, quinces, melons, tomatoes, apricots, and grapes ripen easily.

At Toronto in 1876 the spring and summer rainfall was about 12 inches [1]; at Winnipeg this rainfall reached

[1] See Hayden and Selwyn's 'North America,' p. 467.

more than 15 inches. This copious moisture is of the greatest importance in quickening vegetation when the temperature is high, and everything shoots up with marvellous rapidity. In the treeless regions of the United States there are excessive droughts, where the artemisia, or sage of the desert, is the only form of vegetation. It would seem as if the Great North-West were meant by nature to be the wheat granary of the continent.

(21) The following are the official tables of crop averages for six years, 1882–1887, in Ontario, compared with those of New York and Ohio in the south [1]:—

	Ontario.	New York.	Ohio.
Fall-Wheat per acre	20 bushels.	14 bushels.	12.8 bushels.
Barley ,,	26.1 ,,	22.6 ,,	24 ,,
Oats ,,	35.7 ,,	28.6 ,,	31.3 ,,

Another table of statistics proves the fertility of the Province of Manitoba:—

Spring Wheat.

Manitoba	40 bushels per acre.
Minnesota (best wheat area of United States)	20 ,, ,,
Wisconsin	12.2 ,, ,,
Pennsylvania	12.3 ,, ,,
Ohio	12.8 ,, ,,

In evidence before the Canadian House of Commons on agricultural matters, it was elicited that no less than 60 bushels of spring wheat had been raised to the acre.

Hay also is very cheap and plentiful, and prairie grass when cut and dried averages 2 tons per acre. What is left out does not rot quickly, but is preserved rather than destroyed by the light dry snow, and the animals find winter sustenance by pawing away the snow and eating it.

The northern limits for grains and grasses reach a

---

[1] See p. 18 of 'Agricultural Canada,' by Professor Fream. Published under the direction of the Canadian Government, 1889.

very high latitude in the Canadian Dominion. Rye, barley, and roots can be grown nearly as far north as latitude 65°. The whole available area for grasses and coarser grains is calculated at 2,300,000 square miles. Maize, a tropical plant, sometimes ripens at latitude 54° on the Saskatchewan. At Vermillion (latitude 53° 24″) every kind of garden stuff can be grown. Barley sown on the 8th of May is ready to be cut within three months' time. At Fort Chippewyan, at the entrance to Lake Athabasca, good samples of wheat and barley can be grown. At Fort Simpson, in latitude 61° N., barley always ripens, and wheat is sure four times out of five. At Fort Liard in latitude 61° N. there is the warmest summer temperature in the whole region, and it is said by those who live on the Yucon that barley sometimes ripens under the Arctic Circle in longitude 143° W.[1] At the Forks of the Athabasca, garden produce will grow well, and it was here in 1787 that the explorer Mackenzie found a garden cultivated by a European, who grew all kinds of European vegetables. These are facts elicited from Professor Macoun by a Committee on Immigration to Canada. The following assertion also on the climate was then made: 'It will be seen that about the 20th of April ploughing can commence on Peace River, and from data in my possession, the same may be said of the Saskatchewan region generally. It is a curious fact that spring seems to advance from north-west to south-east at a rate of about 250 miles a day, and in the fall winter begins in Manitoba first and goes westward at the same rate.'

(22) The summer isothermal of 70° crosses Long Island in latitude 41°, passes Chicago at 42°, and then rises on the Saskatchewan to latitude 52° in longitude

---

[1] See p. 83 of 'Guide Book for Settlers.' Published at Ottawa by the Government of Canada, 1882.

110°. It sinks again at the desert areas of the United States to latitude 35° in longitude 105°, and rises to latitude 47° in Oregon. The isothermal of 65° which, on the Atlantic coast, is off Boston, in latitude 42°, rises through Canada to north of Quebec, crosses the Red River at latitude 50° on the 97th meridian, and Mackenzie River near the 60th parallel. This gradual advance northward of the isothermal line in proceeding from east to west, is a fact of the utmost significance with reference to the vegetation of the country.

The following table represents the mean temperature of Toronto and Winnipeg :—

	Toronto.	Winnipeg.
August	66.38	67.34
September	58.18	52.18
October	45.84	35.84
November	36.06	30.66
December	25.78	11.97
January	22.80	6.10
February	22.74	12.32
March	28.93	14.14
April	40.72	39.10
May	51.74	53.13
June	61.85	63.20
July	67.49	68.19

This second table illustrates the average July to August temperature, fit for ripening cereals, at various latitudes [1] :—

	Lat. N.	Summer.	Spring.	Autumn.	July and August.
Cumberland House	53.37	62°	33°	32°	64°
Fort Simpson	61.51	59°	26°	27°	62°
Fort Chippewyan	58.42	58°	22°	31°	60°
Fort William	48.24	59°	39°	37°	60°
Montreal	45.31	67°	39°	45°	68°
Toronto	43.40	64°	42°	46°	66°
Temiscamingue	47.19	65°	37°	40°	66°
Halifax	44.39	61°	31°	46°	66°

[1] See p. 83 of 'Guide Book for Settlers,' 1882.

It will be noticed that although there are many degrees of latitude between Fort Simpson and Toronto, there is, according to the table, a difference of only four degrees of heat during the ripening months of July and August.

(23) **The Forests.**—There is a great variety of trees in the Canadian forests, some of very large dimensions. The black walnut (*juglans nigra*) has an average height of 120 feet, the chestnut attains a height of 100 feet[1]. The climate of the Dominion is a peculiarly suitable one for large trees. They require a large rainfall and a high summer temperature as well, of 65° to 67° F. Such a tree as the sugar-maple will not flourish in Europe, the summer temperature being too low. The deciduous trees are the most numerous in Canada. Of the 114 known species of pines, twenty-one are natives of Canada or of the Hudson's Bay Territory. Alluding to their evergreen appearance in the midst of the Canadian winter, Humboldt has said, 'They proclaim to the inhabitants of the northern regions that, although snow and ice cover the land, the internal life of the plants, like the fire of Prometheus, is never extinguished.' The *Pinus balsamea*, or balm-of-Gilead-fir, and the spruce-hemlock are very beautiful trees. The *Pinus alba* or white-spruce is well known, with its height of 140 feet and long feathering branches. The king of the pine forests is the *Pinus strobus* or Weymouth-pine, so called from the attention given to its cultivation in England by Lord Weymouth. This tree attains the height of 200 feet. 'Some attain a green old age, vigorous to the last, but are prostrated suddenly by the storm which has swept harmlessly over younger heads. Others that have outlived the eagle, sheltered from their earliest youth in some sequestered glade, but now tottering to

---

[1] See Article in the 'Quarterly Review,' No. 217.

their fall, stand bald, spectral, and desolate, waiting only for

> "Some casual shout that breaks the silent air,
> Or the unimaginable touch of Time,"

to bow their heads to the earth. Then geraniums, honeysuckles, wood-lilies, fox-gloves, and fine flowers shoot up around them, and cover for a short time the prostrate trunks with a gorgeous pall, while they collapse and crumble into dust. The tints of the autumnal woods have always excited the astonishment and enthusiasm of travellers. Even in cloudy days the hue of the foliage is at times of so intense a yellow that the light thrown from the trees creates the impression of bright sunshine. Each leaf presents a point of sparkling gold. But the colours of the leafy landscape change and intermingle from day to day, until pink, vermilion, purple, deep indigo and brown, present a combination of beauty that must be seen to be realised, for no artist has yet been able to represent, nor can the imagination picture to itself, the gorgeous spectacle.' The exports of the produce of the forests are worth to the Canadians more than double the combined value of the exported produce of the fisheries and mines[1]. In 1888

The produce of the Mines	was	$4,110,000	(£822,000).
,, ,, Fisheries	,,	$7,793,000	(£1,558,000).
,, ,, Forests	,,	$24,719,000	(£4,943,000).

The following is a description of the Canadian climate during the various seasons :—

'In the summer the excessive heat—the violent paroxysms of thunder—the parching drought—the occasional deluges of rain—the sight of bright-red, bright-blue and other gaudy-plumaged birds—of the brilliant humming-bird, and of innumerable fire-flies that at night

---

[1] See 'Official Handbook of Information,' 1890.

appear like the reflection upon earth of the stars shining above them in the heavens, would almost make the emigrant believe that he was in the tropics.

'As autumn approaches, the various trees of the forests assume hues of every shade of red, yellow and brown, of the most vivid description. The air gradually becomes a healthy and delightful mixture of sunshine and frost, and the golden sunsets are so many glorious assemblages of clouds—some like mountains of white wool, others of the darkest hues—and of broad rays of yellow, of crimson, and of golden light, which, without intermixing, radiate upwards to a great height from the point of the horizon in which the deep red luminary is about to disappear [1].'

As winter approaches the birds migrate south; first the humming-birds, then the pigeons, and next the geese, wild ducks, and plovers, all returning from their breeding haunts in the far north. Those descriptions of scenery and climate painted so often and so vividly by our Arctic explorers along the Polar Seas do not apply to the cultivable zones of grains and grasses further south. The eternal winter night, and the brilliant coruscations of the Aurora Borealis, are reserved for those who adventure deep into the realms of the Ice King. In the latitudes of 50–60° winter is shorn of the extreme terrors of the Arctic Circle, and the colonist enjoying the same amount of daylight as in England, forms for himself regular occupations and regular amusements. The cold, though severe, is dry and conducive to health. In some places cattle graze out all winter, and early in April ploughing begins, and both seeding and ploughing can go on together.

(24) The climate of British North America is determined by great physical causes. The greater extremes in the interior can be explained by the fact that a 'continental'

---

[1] Extract from 'The Emigrant,' by Sir F. B. Head.

as opposed to a 'maritime' or 'insular' climate is always subject to great variations. The ocean is able to store heat for a longer time than land, and on the sea-coast or on islands temperature is more uniform. In England, surrounded by water, our average temperature ranges from 40° to 60°, in corresponding latitudes in Central Asia the average is from 0° in winter to about 70° in summer. The ocean currents also play a great part in determining the climate of the Atlantic and Pacific coasts of the Dominion. The coasts of Labrador and Newfoundland and the Gulf of St. Lawrence, are chilled and frozen by the icy current down from Davis's Straits and Baffin's Bay, bringing with it floes and icebergs down into the Atlantic almost to the latitude of Malta. The Gulf Stream, escaping from the Gulf of Mexico, just misses Newfoundland, and flows in a north-easterly direction past the British Isles into the Arctic Gulf, between Spitzbergen and Nova Zembla. It thus happens that the water which reaches the North Cape of Norway in 71°, is of nearly the same temperature as that of the harbour of New York on the opposite side of the Atlantic in latitude 40°.

The vast expanses of the North American lakes have an effect upon the climate and keep it more uniform. The waters of the deeper lakes do not freeze in the winter except for a few miles out from shore, because the storage of heat gained during the summer months does not quickly escape from these immense inland seas of vast depth. In a small way they perform, as at Toronto, the functions of the seas, and reproduce, if we may so term it, an inland maritime climate on a modified scale.

As the Canadian Dominion is a land of forests, the trees make a great difference to the climate. If the surface of a land is stripped of trees and herbage by fires or other destructive agencies, it becomes more quickly heated by the sun, and evaporation goes on more speedily. The immense forests of North America have materially assisted the rainfall in the country. At the same time they have contributed to its richness. For centuries the rotting leaves and foliage from deciduous trees have been accumulating, until the surface of the earth is covered with a fruitful and reproductive soil which the axe of the lumberer discloses to the light.

The action of snow upon the land is beneficial. It acts as a covering and protection, and prevents the earth being frozen down to any great depth. As a rule the earth in Canada is not frozen to a greater depth than twelve or eighteen inches. When the snow melts beneath the thaws of spring, the husbandman, especially in Manitoba, finds the fertile soil which he has ploughed in the autumn easily worked, thus saving him much labour. From another point of view the ice and snow are beneficial. For purposes of traffic and communication they open up the whole country. King Frost is the great Macadam of the country, providing an easy way for wagon and sleigh through the most inaccessible districts. The lumberer carries on his most active operations in the winter months.

The question of soils is most important to husbandmen who have to make their living on them and from them. The extreme richness and variety[1] of the Prairie soils have only recently been fully known, as the resources of the North-West have been more scientifically analysed. Moreover, in the absence of agues and fevers these soils are especially healthy to live upon.

(25) Dr. Philpot, Surgeon to Her Majesty's Forces, 1871, writes as follows:—'Canada is an exceptionally healthy country. I do not hesitate to make the statement after seven years in the country, engaged in an

[1] Appendix IV.

extensive medical practice. As a race, the Canadians are fine, tall, handsome, and powerful men; well built, active, tough as pine-knot and bearded like pards. The good food upon which they have been brought up (with the invigorating climate), appears to develop them to the fullest proportion of the "genus homo".'

On the sea-coast and along the shores of Newfoundland and Nova Scotia a hardy population of sailors and fishermen have been reproducing for generations the skill and daring of the old stock: if there were a Transatlantic wing of an Imperial fleet, here would be the material to recruit from. The Newfoundland 'Jack Tar' is as loyal as his compatriot at home, and perhaps he has more skill and experience in practical arts of seamanship. His is a coast that tests to the uttermost his native courage. Although chilly and extremely rigorous at times, the climate of Canada is not so trying, by sea or land, as that of England, and it gives a new vigour. There is a discernible difference between the Canadian and American types, and it is thought that the northerners contrast very favourably with the southerners. They are living as a nation in a home which is more like that of Northern Europe, from which they sprang. In the valley of the St. Lawrence we may expect a hardier stock than in the valley of the Mississippi. The low readings of the thermometer during the Canadian winters, which at first sight seem extremely prejudicial to health, are not really so. The Prairie air is extremely dry, and 'enveloping the body in a medium, conserves its animal warmth, offering no facility for escape[1].' The frames, therefore, of the Canadian colonists are hardy and robust. Their national courage remains the same, prompting to great enterprises.

On more than one occasion Canadians have desired to

[1] See p. 33 of 'Agricultural Canada,' 1889.

share the perils and burdens of the mother-country. The most recent and perhaps the most notable instance of this was during the Soudan campaign, when a number of Canadian *royageurs* took part in the Nile Expedition under Lord Wolseley.

(26) **The Canadian army.** — The Canadian Reserve nominally consists of all male British subjects between eighteen and sixty, numbering about 700,000 men[1].

The following was the disposition of the active force, January 1, 1888:—

Cavalry	1,987
Field Artillery	1,440
Garrison Artillery	3,479
Engineers	179
Infantry and Rifles	31,506
	38,591

The Dominion is divided into twelve military districts:—

Ontario	4
Quebec	3
Nova Scotia	1
New Brunswick	1
Prince Edward Island	1
British Columbia	1
Manitoba	1
	12

The different provinces of the Canadian Dominion have, in each case, their geographical and climatic peculiarities, as may be gathered from a study of the general features of the continent. The maritime provinces of Nova Scotia, New Brunswick and Prince Edward Island differ from the river and lake provinces of Quebec and Ontario. Manitoba and the districts of the prairies naturally differ from both, and when we cross the Rocky Mountains we find in the province of British Columbia

---

[1] See Hayden and Selwyn's 'North America,' p 542.

distinctive and peculiar features. The range of country is vast, running for a distance of 3000 miles through 75° of longitude, i. e. from 55° to 130°, from Newfoundland to Queen Charlotte's Island. At each step westward we are brought face to face with new revelations of nature on a grand and colossal scale. The sky is bluer, the mountains higher, the rivers larger, the forests bigger, the plains broader, than in the Old World.

CHAPTER II.

Quebec.

(1) THE Province of *Quebec*, the first important centre of French colonisation, is traversed by the St. Lawrence, and has the advantages both of a maritime and a continental country. It extends from east to west between 57° 50′ and 80° 6′ W. longitude and from south to north between 52° and 45° N. latitude[1]. Its shape resembles that of a triangle with its base to the south-west and its apex at L'Anse au Sablon, just inside the Straits of Belle Isle. Its greatest length from the north-west corner to the Straits of Belle Isle is about 1350 miles; its greatest width along the 71st and 72nd degrees of W. longitude about 500 miles. It is bounded on the

---

[1] See p. 7 of 'General Sketch of the Province of Quebec,' by the Hon. Honoré Mercier, Premier of the Province, 1889.

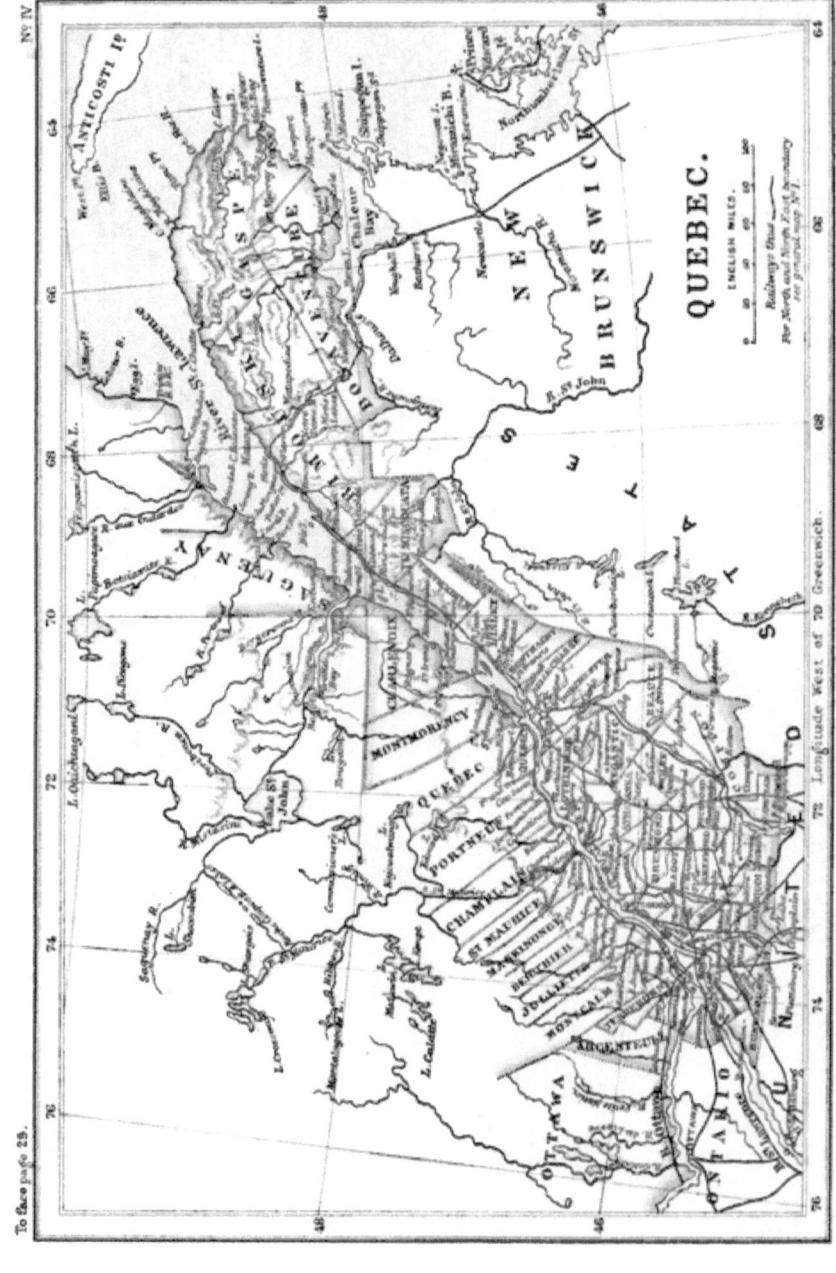

east, south-east, and south by the Gulf of St. Lawrence, the Bay des Chaleurs, the river Restigouche, and the interprovincial line which divides it from New Brunswick, thence by the international line dividing Canada from the United States to the *Hall River*; thence by the 45th degree of N. latitude to its intersection with the middle of the river St. Lawrence at *Point St. Regis*; to the south-west, west, and north-west by the middle of the river St. Lawrence from *Point St. Regis* to *Point à Beaudet*; thence by the interprovincial line which separates it from Ontario to *Point Fortune* on the Ottawa; thence along the middle of the Ottawa and *Lake Temiscamingue* to the northern extremity of that lake; thence by a meridian line to James' Bay; to the north-west and north by James' Bay as far as the mouth of the East Main River[1], by the right shore of this river from its mouth to its source; thence going north by a line striking the most northern waters of the great river Esquimaux, thence in a circular line till the 52nd degree N. latitude is struck, and thence to the St. Lawrence River.

(2) The area of this Province is 258,634 square miles. Deducting the surface of the inland waters and those of the River and Gulf of St. Lawrence, the land amounts to 188,688 square miles or 120,764,651 acres. The whole area exceeds that of France by 54,000 square miles. The perimeter of the whole province is about 3000 miles, of which 740 miles are sea-coast and 2260 miles land-frontier. If we count the north and south shores of the St. Lawrence as an interior development of coast along which steamers can pass we get an extra distance of 1150 miles. The Islands of Anticosti and Brion, the Bird Rocks, the Magdalen Islands, a barren group about fifty miles north of Prince Edward Island, and all the islands near Gaspé

---

[1] See p. 8 of Honoré Mercier's 'Sketch of the Province of Quebec.'

belong to the province of Quebec. Anticosti is an uncultivated island about 145 miles long and 30 miles broad, dividing the St. Lawrence Gulf into two channels. It is chiefly used as a fishing station in the summer. Montreal Island and the Isle of Orleans are in the St. Lawrence, Allumet and Calumet in the Ottawa [1].

The province of Quebec is chiefly made up of that great basin, whose waters flow towards the St. Lawrence. The valley of the Richelieu forms a kind of inner triangle with its head at the entrance of Lake St. Peter and its base at the international boundary on the south. The area is 1400 square miles, and the surface is level with the exception of a few eminences. From one of these eminences called 'Belœil,' about 1400 feet high, a magnificent view of this fertile plain can be taken in. To the north Montreal and the great Victoria Bridge can be seen, to the south Lake Champlain, fifty miles distant.

(3) **Mountains.**—Quebec has two mountain systems. (i.) The *Laurentian* with a general trend from north-east to south-west. From Labrador on the east to the neighbourhood of the Saguenay these mountains form a compact barrier only broken through by the large rivers which cut them transversely. In approaching the Saguenay the chain separates into two distinct ranges: (1) that of the 'height of lands' which describes a great curve to the north in the direction of Lake St. John; (2) that of the Laurentides, properly so called, skirting the north bank of the St. Lawrence River and receding gradually from it to a distance of thirty miles in rear of Montreal, forming the southern watershed of the basin of Lake St. John and the Ottawa River. Along this whole distance the average height is about 1600 feet. Near Lake St. John the principal crest of the mountains is 4000 feet.

[1] See Honoré Mercier's 'Sketch of the Province of Quebec.'

Cape Tourmente, just below Quebec, is 1919 feet; St. Anne's Mount, twenty miles below, 2687 feet. It is the presence of these mountains which adds to the romantic beauty of Quebec. In the county of Argenteuil, north of the Ottawa River, the highest summit is that of 'the Trembling Mountain,' 2060 feet. The average elevation of the interior plateau of the province in which the basin of Lake St. John and the Upper Ottawa is situated is about 600 feet above the sea. Lake St. John is 293 feet above the sea. Amongst these mountains there are dense forests of conifers and hard woods. The valleys especially abound in pine, spruce, and cedar.

To the geologists, the Laurentian Rocks, extending over an area larger than France, have been extremely interesting as throwing light upon their science and opening out an older leaf of the world's history than any hitherto known. Amongst the limestone beds of the Laurentian range a fossil was discovered and named the *Eozoon Canadense*, the oldest relic of life which has been found upon the globe.

'The Laurentian Rocks must have been separated by a vast lapse of time from the next formation which succeeds them. For during that interval they had been changed from the state of sand, mud, and gravel, into gnarled crystalline gneiss, schist, and quartz-rock, and in that altered state had been anew exposed to denudation. It is beyond that immense gap that Sir William Logan's discovery enables us to throw back the beginning of life. . . . Whilst chronicling this Canadian discovery— which marks an epoch in the history of geology—Sir Roderick Murchison has shown that a representative of the Laurentian rocks exists in Great Britain, and the whole of the Scottish Highlands has been brought into relation with the rocks of the rest of the island[1].'

[1] 'Quarterly Review,' No. 249. 'Siluria,' by Sir Roderick Murchison.

(ii.) The *Alleghany system*. This chain is simply a prolongation of the Appalachians. Beginning from the east end of the Gaspé Peninsula they skirt the southern shores of the St. Lawrence, and only begin to trend away from it at Kamouraska. The first principal axis bends towards the river and runs to the north-west in the direction of St. Ann's Mount, and then inclines towards the south-west to form the heights between the Cape Chatte and Matane Rivers, diverges in the direction of the Chaudière River opposite Quebec, beyond which the principal ridge runs south-west and then southward into Vermont territory, where it is known as the Green Mountain. From Gaspé to Quebec this mountain chain forms the watershed between the basin of the St. Lawrence to the north and the Bay des Chaleurs and Bay of Fundy on the south. It gives a very imposing appearance, especially at Cape Chatte, to the St. Lawrence River. Some of the peaks in the region between St. Anne and Matane Rivers attain a lofty height. Mount Logan is 3768 feet high. These mountains are known generally as the Notre Dame Range.

(4) **Lakes.**—The small lakes of Quebec province are very numerous and can be counted in hundreds, especially on the north side of the St. Lawrence amongst the Laurentian Mountains, between the Ottawa and Saguenay Rivers, but as a rule they are comparatively small. Lake St. John has a superficies of 360 miles; *Grand Lake*, latitude 48 and longitude 77 W., one of 550 miles. Lake Crossways lies at the sources of the St. Maurice, Lakes Kempt, Manouan, and La Collotte form a group further south. Lake *Temiscamingue* lies at the sources of the Ottawa, Lake Shecoubish at those of the Saguenay, Mistassinnie at those of the Rupert. Lakes Edward and Kajoualwang are south of St. John's Lake. Lake St. Peter is formed by the St. Lawrence.

(5) **Rivers.**—The St. Lawrence rises in a small lake in Minnesota, which discharges its waters into Lake Superior by the River St. Louis. It is designated by different names; St. Mary's between Lake Superior and Lake Huron; St. Clair or Detroit between Lake Huron and Lake Erie; Niagara between Lake Erie and Lake Ontario; and lastly, St. Lawrence, from the latter lake to Pointe des Monts, which is regarded as the line of separation between the river and the gulf. 'The total length of the St. Lawrence is 2180 miles. Its ordinary width varies between one and four miles in its upper course, increasing below Quebec to twenty and thirty miles at its mouth. It is navigable for ocean vessels to Montreal, which is 833 miles from the Straits of Belle Isle; and from Montreal to the head of Lake Superior, a distance of 1398 miles, can be navigated by vessels of 700 tons with the aid of canals built to overcome the rapids. The smallest locks of these canals are 270 feet long, 45 wide, with 9 feet of water[1].' The tidal current is felt on the St. Lawrence as far as Three Rivers, half-way between Montreal and Quebec.

The *Ottawa* is the largest tributary of the St. Lawrence. It is about 600 miles long, and drains an area of 80,000 square miles. It rises in the west of the province, and forms a boundary line between Quebec and Ontario for 400 miles. It is navigable for 250 miles, the rapids and falls being avoided by canals. The Ottawa is a picturesque river, and is fringed by the noblest forests in the world. It is on the Ottawa and its tributaries that the lumberer finds the best timber. The Ottawa is described as 'a long succession of reaches studded with islands, narrow passes, fair lakes, impetuous rapids, and magnificent falls. The voyage on its water, day after day, is a succession of charming surprises. At one time a wide prospect of

---

[1] See p. 43 of Honoré Mercier's 'Sketch of the Province of Quebec.'

open lake reaches almost to the horizon; at another, you look over an endless undulating extent of hill and dale; then you are shut up within a narrow gorge without visible escape. To increase the feeling of exhilaration which variety gives, the traveller is compelled to change perpetually his mode of conveyance—from steam to stage, from boat to ferry, from car to scow. At some intervals the traveller may have to walk; for instance, where a narrow platform, thirty feet high, crossing a rocky valley, has been burnt by a fire in the woods.' The tributaries of the Ottawa are—

River du Moine [1]	. . .	80 miles
,, Noire	. . .	115 ,,
,, Gâtineau	. . .	250 ,,
,, du Lièvre	. . .	170 ,,
,, Petite Nation	. .	50 ,,
,, Rouge	. . .	120 ,,
,, du Nord	. . .	60 ,,

The *Saguenay* is a tributary of the St. Lawrence, and flows from Lake St. John. It is a stern and gloomy river of immense depth, and has been described as 'a tremendous chasm cleft in a nearly straight line for some sixty miles.' It brings down the waters of Lake St. John, which receives the inflow of fourteen streams. Like the Ottawa, it boasts of magnificent scenery and forests. Thirty-five miles from its mouth are two lofty capes, some 2000 feet high, called Capes Trinity and Eternity. At the mouth of the Saguenay is Tadousac, 134 miles from Quebec, one of the first posts occupied by the French with the purpose of securing the fur-trade. Some distance up the river is Chicoutimi, an Indian word meaning 'deep water,' where the navigation ends.

The *Montmorency* joins the St. Lawrence 8 miles below Quebec. It is noted for its famous Falls, where a stream of water 50 feet wide rushes over a steep precipice 250

---

[1] See p. 15 of Honoré Mercier's 'Sketch of the Province of Quebec.'

feet high. It is one of the many 'sights' of Canada, and its appearance in winter when a solid dome of ice 200 feet high is piled up is wonderful. These Falls, being near Quebec, are often visited by 'sleigh' and 'toboggin' parties. Close by is Haldimand House, where the Duke of Kent, Queen Victoria's father, once lived.

Other important rivers *on the north side* of the St. Lawrence are :—

	Length.
St. Maurice	280 miles
Batiscan	93 ,,
Portneuf	80 ,,
Betsiamite	112 ,,
Outardes	234 ,,
Manicouagan	224 ,,
St. John's	150 ,,
Des Esquimaux	100 ,,

On the south side the chief rivers are :—

The *Richelieu*, flowing from Lake Champlain on the south side into the St. Lawrence. It has a course of 50 miles.

The *St. Francis* rises in Lake Memphramagog, and after a course of 100 miles flows into Lake St. Peter. The Yamaska is about the same length.

The *Chaudière* flows from Lake Megantic, and enters the St. Lawrence nearly opposite Quebec. This river is noted for its beautiful Falls.

(6) **Counties and Districts.**—The province of Quebec is divided into sixty counties :—

1. Argenteuil.
2. Bagot.
3. Beauce.
4. Beauharnois.
5. Berthier.
6. Bonaventure.
7. Belle Chasse.
8. Brome.
9. Chambly.
10. Champlain.
11. Charlevoix.
12. Châteauguay.
13. { Chicoutimi. / Saguenay. }
14. Compton.
15. Dorchester.
16. { Drummond. / Athabasca. }

17. Gaspé.
18. Hochelaga.
19. Huntingdon.
20. Iberville.
21. Jacques Cartier.
22. Joliette.
23. Kamouraska.
24. La Prairie.
25. L'Assomption.
26. Laval.
27. Levis.
28. L'Islet.
29. Lotbinière.
30. Maskinonge.
31. Megantic.
32. Missisiquoi.
33. Montcalm.
34. Montmagny.
35. Montmorency.
36. { Montreal, C.
       ,, E.
       ,, W. }
37. Napierville.
38. Nicolet.
39. Ottawa.
40. Pontiac.
41. Portneuf.
42. { Quebec, C.
       ,, E.
       ,, W. }
    Quebec Co.
43. { Richmond.
      Wolfe. }
44. Richelieu.
45. Rimouski.
46. Rouville.
47. St. Hyacinthe.
48. St John's.
49. St. Maurice.
50. Shefford.
51. Sherbrooke.
52. Soulanges.
53. Stanstead.
54. Temiscouata.
55. Terrebonne
56. Three Rivers.
57. Two Mountains.
58. Vaudreuil.
59. Verchères.
60. Yamaska.

In these 60 counties there are 65 electoral districts, the difference being caused by Montreal having 3, and Quebec 4 electoral districts.

The total area of these counties is calculated to be 130,000,000 acres. This is distributed as follows:—

	Acres.
Conceded in fiefs	10,678,981
In full and common socage (townships)	8,950,953
Surveyed in Town Lots	6,400,000
Awaiting Survey	103,970,066
	130,000,000

(7) **Towns and Cities.**— Quebec has an historical interest of its own, which makes it different from all others. The French language, laws, and customs, are perpetuated amongst its inhabitants. A recapitulation of the names of its counties and towns alone would be a proof of its history. Such names as Champlain, Jacques Cartier, Montcalm, Montmorency, Richelieu, St. Hyacinthe, St. Maurice, St. John's, Vaudreuil, recall vividly the chief epochs of the history of what was once called New France. Of the whole population 80 per cent. are of French extraction [1].

Each county has a county-town, and next to Montreal and Quebec the most important are :—Richmond, the county-town of Richmond ; Sherbrooke, of Sherbrooke ; Three Rivers, of Three Rivers ; St. John's, of St. John's ; St. Hyacinthe, of St. Hyacinthe.

Quebec (65,000) is the provincial capital of the immense province of Quebec. It is a very old and interesting city. In 1608, Samuel Champlain, the great French explorer, founded Quebec at the confluence of the St. Lawrence and St. Charles Rivers. It occupies a most magnificent site, and was formerly so strongly fortified that it was deemed to be impregnable. Here, on the 'Plains of Abraham,' the decisive battle for the possession of Canada was fought in 1759, and here Wolfe and Montcalm fell. It is 360 miles from the mouth of the St. Lawrence, and 180 miles from Montreal.

'Let the spectator stand on the Flagstaff Battery within the lines won by Wolfe's gallantry, but which he could not live to enter. Below lie the steep narrow streets of a city as French as Havre or Calais. Yonder is an open market-place, with groups of women sitting at their stalls with kerchiefed heads. At a distant corner-house you may see a shrine to Our Lady, newly white-

---

[1] Honoré Mercier's 'Sketch of the Province of Quebec.'

washed by the piety of the inmates. The first stone block adjoining the Cathedral is the celebrated Roman Catholic College, the Laval University, named in honour of the first Bishop of Quebec. Another great pile of stone is the Parliament House for the provincial legislature. . . . Then, still below, the shores are lined with warehouses and quays and masses of shipping. All the surrounding waters are filled with sails; the scene is one of sunlight and life. Steamers with their filmy lines of smoke pass up and down the river, or rapidly across. At Point Levis, opposite the citadel, lie stranded or lazily floating incalculable masses of timber, waiting for transit to the British Isles, South America, or Australia.

'Quebec sits on her impregnable heights a queen amongst the cities of the New World. At her feet flows the noble St. Lawrence, the fit highway into a great empire, here narrowed to a mile's breadth, though lower down the water widens to a score of miles, and at the Gulf to a hundred. From the compression of the great river at this spot the city derives its name, the word signifying in the native Indian tongue a strait. On the east of the city, along a richly fertile valley, flows the beautiful St. Charles, to join its waters with that of the great river. The mingled waters divide to clasp the fair and fertile "Isle of Dreams."

'The city, as seen from a distance, rises stately and solemn, like a grand pile of monumental buildings, clustering houses, tall, irregular, with high-pitched roofs crowd the long line of shore and climb the rocky heights. Great piles of stone churches, colleges, and public buildings, crowned with gleaming minarets, rise above the mass of dwellings. The clear air permits of the use of tin for the roofs and spires, and the dark stonework is relieved with gleaming light. Above all rise the long dark lines of one of the world's famous citadels[1].'

(8) *Montreal* has a population of more than 200,000. It stands on Montreal Island, and was founded on the site of Hochelaga in 1642. This is a description by Parkman of the first founding of Montreal. It was in its beginning a religious enterprise, conducted by enthusiastic men and women of France. 'Maisonneuve sprang ashore and fell on his knees. His followers imitated his example; and all joined their voices in enthusiastic songs of thanksgiving. Tents, baggage, arms, and stores were landed. An altar was raised on a pleasant spot close at hand, and Mademoiselle Nance, with Madame de la Peltrie, aided by her servant Charlotte Barré, decorated it with a taste which won the admiration of the beholders. Now all the company gathered before the shrine. Here stood Vimont in the rich vestments of his office. Here too were the two ladies with their servant, here Montmagny, no very willing spectator, and Maisonneuve, a warlike figure tall and erect, his men clustering round him. They kneeled in reverent silence as the Host was raised aloft, and when the rite was over the priest turned and addressed them. "You are a grain of mustard seed that shall rise and grow till its branches overshadow the earth. You are few, but your work is the work of God. His smile is on you, and your children shall fill the land."'

In addition to the above may be mentioned the following cities and towns[2]:—Levis, 12,175; Hull, 6,890; Sorel, 5,791; Valley Field, 3,906; Nicolet, 3,764; Joliette, 3,268; Lachine, 2,406; Longueil, 2,335; Frazerville, 2,291; St. Jérôme, 2,032; Chicoutimi, 1,935; Farnham, 1,880; Iberville, 1,847; Beauharnois, 1,499;

---

[1] Marshall's 'Canadian Dominion.'
[2] Mercier's 'Handbook,' 1890.

Rimouski, 1,417; Terrebonne, 1,398; Louisville, 1,381; L'Assomption, 1,313; Berthier, 1,039. According to the statement of the Honourable Honoré Mercier, Premier of Quebec, the rural population constitutes 76.29 per cent. He also observes that from 1871–1881 the Roman Catholic population showed an increase of 150,866 or 14.79 per cent. Supposing the progression to continue in the same ratio, the present decade will give an increase of 173,149, which will bring the number of Roman Catholics in 1891 to 1,343,867, or 87.97 per cent. of the total population of the Province.

As regards callings the Census of 1881 grouped the Quebec population as follows:—

Agricultural . . .	201,963 or	48.63 per cent.
Industrial . . .	81,643 ,,	19.67 ,,
Commercial . . .	34,346 ,,	8.27 ,,
Domestic . . .	24,267 ,,	5.85 ,,
Unclassified . . .	72,635 ,,	17.50 ,,

(9) The *Eastern Townships* are a peculiar feature of the province of Quebec. They lie close to Vermont and the United States frontier on the parallel of 45° N., in the vicinity of Lakes Memphramagog, Megantic, and Massawippi, which are of surpassing beauty. The soil is very fertile, and the forests are full of timber. It is a purely agricultural district. These townships were originally settled by United Empire Loyalists, who adhered to England at the time of the American War. It is regarded as the 'English' portion of the province of Quebec.

(10) There are five main centres of colonisation in the province of Quebec[1]: (1) The Valley of the Saguenay, where the extent of disposable land is 616,600 acres, valued at 10d. per acre; (2) The Valley of St. Maurice,

---

[1] See p. 140 of Silver's 'Handbook to Canada' (1881).

with 440,000 acres, for sale at 1s. 3d. per acre; (3) The Valley of the Ottawa, with 1,358,000 acres, at 1s. 3d. per acre; (4) The Eastern Townships, with 850,000 acres, at 2s. 1d. or 2s. 6d. per acre; (5) Gaspé, where there are 491,100 acres, at the rate of 10d. to 1s. 3d. per acre. Besides these centres, there are 1½ million acres disposable in the valley of the Lower St. Lawrence. An emigrant, therefore, may find abundance of room in the province of Quebec which, although it is the oldest of all, does not hold on the average six persons to the square mile. Elsewhere in the Dominion the space is absolutely unlimited.

(11) The shipping of the province of Quebec in 1886 was as follows:—

	Ships.	Tonnage.
Amherst	33	1,092
Gaspé	44	2,347
Montreal	1,007	136,286
Percé	3	133
Quebec	856	101,481
	1,975	232,556

This is exclusive of the fishing fleet[1]. At the close of 1885 more than 160 vessels were afloat, and 7,949 boats giving employment to nearly 11,322 men.

In 1888 the total shipping of the Canadian Dominion was 1,089,642 tons.

[1] See Biggar's 'Canada,' 1889.

CHAPTER III.

Ontario.

(1) WHEN Champlain first explored the St. Lawrence, beyond Montreal, he came to a lake which for beauty and size seemed to surpass all others. The Indians called it Ontario, meaning in their language 'The Beautiful Water,' and the explorer accepted the name. The Province of Ontario is the province of great lakes, as Quebec is the province of the great river which drains them. Formerly Ontario was called Upper Canada, and it received its present name and meaning in 1867. In contrast again with Quebec, Ontario is almost a purely British colony in population, customs, and traditions. Quebec and Ontario, although they are near neighbours occupying conterminous territories, are to all intents and purposes distinct French and British colonies. That they should be so completely in accord on common subjects, and so loyal to a common centre of government, proves the power of Federal principles.

Area.—The area of Ontario is calculated to be about 181,800 square miles[1], one and a half times the size of Great Britain and Ireland. No less than 25,000,000 acres have been surveyed. The boundaries of the Province are the River Ottawa and Quebec on the northeast, the St. Lawrence and the Lakes of Ontario and Erie on the south. On the west, it is bounded by Manitoba and the North-West Territories. The Western Peninsula,

[1] See p. 7 of 'Official Handbook,' 1890.

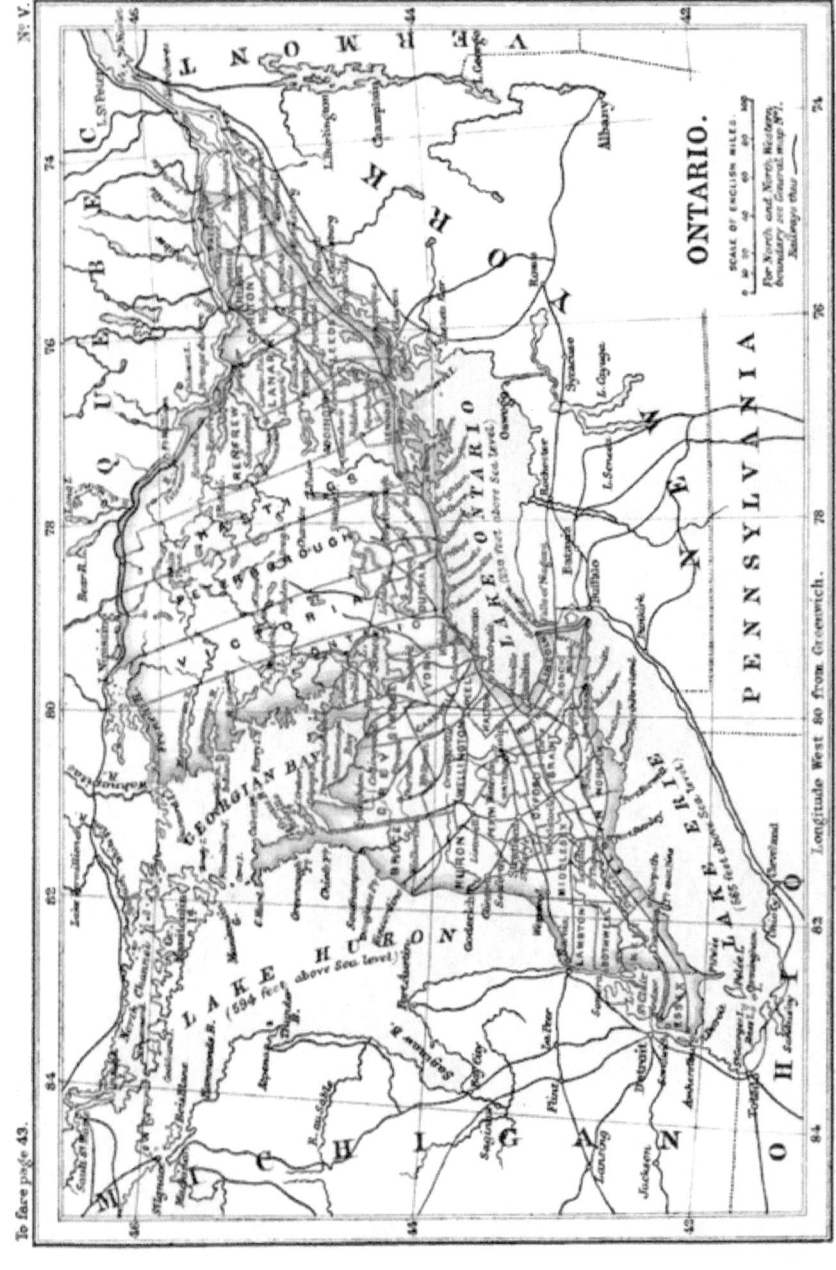

extending from Toronto to Lake St. Clair, has been called the 'Garden of Canada.'

(2) **Surface.**—The highest mountains in Ontario are the Blue Mountains, on the south of Nottawasaga Bay. They do not exceed 1000 feet. As a rule the country is flat or gently undulating, thus differing from the sister province of Quebec.

The physical features and characteristics of the Province of Ontario are comparatively easy to understand. A glance at the map will show that Nature has given it a commanding position. Bordering on United States territory on the south and south-west, and reaching in its furthest north-west limits to the Manitoba lake system, it seems fitted admirably by nature to be a distributing centre for the north and south parts of the continent. The traffic by the Sault Ste. Marie is said already to be equal to the tonnage that passes through the Suez Canal. Its railway system connects with that of the United States at half a dozen different points: the markets throughout the province are within easy range of the farmer in every settled district, the highways are kept in good repair, the towns and villages are thickly dotted over the country, being seldom more than five to ten miles apart, and, excepting in the new and far northern settlements, almost every farm is within fifteen miles of a railway station, *showing a healthy distribution of rural population.*

(3) **Soil and Products.**—The country has many varieties of soil, nearly all of which are fertile and easy of cultivation. The farms yield a good return if cultivated with a view to stock-raising or dairy farming, fruit and mixed farming, the branches which promise in the future to be the leading features of agricultural industry in Ontario. Means of transport are ample, and freights are low.

An Agricultural Return, collected by the Bureau of

Industries for the Province of Ontario, gives the following average production of field crops per acre for the whole Province of Ontario in 1888, together with their yields:—

		Total yield.	Per acre.
Fall-wheat	(bushels)	18,071,142	16.7
Spring	,,	9,518,553	17.5
Barley	,,	19,512,278	26.1
Oats	,,	58,665,608	35.4
Rye	,,	1,106,462	15.4
Peas	,,	16,043,734	20.5
Buckwheat	,,	1,678,708	21.2
Beans	,,	482,072	23.5
Potatoes	,,	16,012,358	144.7
Mangolds	,,	8,787,743	467.0
Carrots	,,	3,478,751	338.3
Turnips	,,	47,061,053	420.9
Hay and clover (tons)		2,994,446	.88 [1]

(4) **Towns.**—The first town that claims attention in Ontario is Ottawa (44,000), the seat of the Federal Government. It is the centre also of the Ontario lumber trade. The Ottawa is the river by which, it will be recollected, the early explorers journeyed to the West across Lake Nipissing and French River, when the Niagara route on the south was closed against them by the hostility of the Iroquois. The city is in easy and quick communication with the St. Lawrence. It is thus described by the Marquis of Lorne in his 'Canadian Pictures':—

'The city is placed on the banks of a broad stream, which narrows at one spot above the town and pours over a steep ledge of rock to expand immediately afterwards to flow on in a channel navigable except at one place where there are rapids, until it empties itself, about 80 miles away, into the St. Lawrence. Forty miles to the south, the last named mighty river is the boundary

[1] Extract from 'Official Handbook,' published by the Canadian Government, Oct. 1890.

between Canada and the State of New York. To the north-west the Ottawa stretches on far into the wilds, having its head-waters at the height of land which divides the basin of the St. Lawrence from that of Hudson's Bay.'

Kingston (17,000) is described by the same writer as one of the pleasantest of Canadian towns. It lies in one of the oldest settled districts of Ontario and is situated on the Cataraqui River, being connected with Ottawa by the Rideau Canal. The Grand Trunk Railway passes through it, and the steamers from Toronto and Montreal call at the port. 'Picturesque Martello towers rise from the waters and are posted along the environs of the town to where Fort Henry, on the hill to the southward, dominates the landscape. The streets of the lime-stone built city are well planted. Ship and boat-building with the several manufactories and the stir at the wharves caused by the transhipment of grain, keep a good deal of life in the locality, deserted as it is by troops and politicians. The traces of the old French fort built by Frontenac are yet visible . . . From Kingston the so-called thousand isles may be seen by taking the steamer down to Montreal . . . The width of the stream near Kingston is about seven miles and the whole area for many miles down is a labyrinthine maze of water, the rocky wood-clad group of islets separating the deep, strong-running channels.'

(5) At the west end of Lake Ontario is *Toronto*, meaning 'the meeting place of the tribes,' once called *York* (172,000). It is the seat of the Provincial Government, and has many important manufactories. Toronto is as near Liverpool, by the St. Lawrence and the Straits of Belle Isle, as New York is by the ordinary sea-route. It is called the 'Queen City of the West' with its 'array of dome and turret, arch and spire, and

the varied movement of its water-frontage is one that cannot fail to evoke pleasure and create surprise ... The city, which covers an area of eight or ten square miles, is built on a low-lying plain with a rising inclination to the upper or northern end, where a ridge bounds it, probably the ancient margin of the lake. Within this area there are close upon 120 miles of streets, laid out after a rigid chess-board pattern, though monotony is avoided by the prevalence of Boulevards and ornamental trees in the streets and avenues. What the city lacks in picturesqueness of situation is atoned for by its beautiful harbour and by its private gardens and public parks. The Custom House, with its adjacent examining warehouse, is perhaps one of the most striking instances of the new architectural régime. The business done here rates the city as second port of entry in the Dominion [1].'

Mr. Marshall remarks in his work on 'The Canadian Dominion,' that 'the University of Toronto is perhaps the only piece of collegiate architecture on the American continent worthy of standing in the streets of Oxford [2].'

Niagara. The Falls of Niagara have been a theme of wonder ever since Father Hennepin first saw them and described them in his travels. No description can do them justice. Sir Charles Lyell calculated that 1,500,000 cubic feet of water pass every minute. The waters are gradually wearing the rock away and the shape of the Falls has changed much since the Jesuit explorer saw them. The height of the American Falls is said to be 164 feet, of the Horseshoe Falls 158 feet. To Hennepin they seemed 600 feet high. It is said that the only Falls to be compared with them are the Victoria Falls of the Zambesi in South Africa. Charles Dickens in his

---

[1] 'Picturesque Canada,' edited by the Very Reverend G. M. Grant, Principal of Queen's University, Kingston, Ontario.

[2] This beautiful structure was destroyed by fire, Feb. 14, 1890.

'American Notes' has pourtrayed the 'wreathing waters in the rapids hurrying to take their plunge,' then 'the giant leap' and the 'rainbows spanning them a hundred feet below,' and that 'tremendous ghost of spray and mist which is never laid, arising from its unfathomable grave.' The Indian says that this rising mist is the 'incense of the world rising to the Great Spirit.'

St. Catharine's (11,000) is situated on Twelve Mile Creek, and is the principal point on the Welland Canal. 'The country in its neighbourhood is like that of a great part of the peninsula between Lakes Erie and Huron, very fertile and originally covered with a fair growth of maple and other hard wood. It has now been carved out into excellent farms, occupied by people mainly Scots and English in descent.'

(6) *London* (27,000) in Middlesex is one of the most enterprising and prosperous of the Canadian cities. It is built on a little stream called the Thames; her bridges are named Blackfriars and Westminster; her principal church is St. Paul's; her streets are Piccadilly, Oxford Street, Regent Street, and Pall Mall. On the east on Lake Erie is Port Dover. Amongst other important towns are *Ingersoll* in the county of Oxford; *Guelph* (11,000) famous for its Agricultural College; *Woodstock* (8,314) in the county of Oxford; *Stratford* (9,000) with an adjacent village of Shakespeare in Perth; *Walkertown* in the county of Bruce on the river Saugeen; *Hamilton* (43,082) near Niagara, sometimes called 'the ambitious' with a desire to rival *Toronto*; *Brantford* (13,000) which is brought into communication with Lake Erie by a canal. The name of Milton is commemorated in County Halton. Along Lake Ontario, on the northern shore, there is a Scarborough and Whitby. There is another Chatham in County Kent, and not far off a Colchester, Rochester, Maidstone, and Windsor, and Sandwich. The very

names of the counties and towns of Ontario betray their English origin.

A journey by steamer from Montreal to the extreme end of the great Province of Ontario is thus sketched by the Marquis of Lorne:—

'But here (at Montreal) if she be a ship of small tonnage her journey need not be terminated. Rapid waters flash over the rocky ledges in the stream above, and the continuation of these rapids, which are often almost cascades, bars her direct progress; but at these she finds magnificent canals constructed with 9 and sometimes 14 feet of water over the sills of the locks, and she can proceed until the majestic waters of Lake Ontario allow her again for 150 miles to proceed upon her course. Then, when the steam of the Falls of Niagara rises above the plains which seem to shut out further advance, she slips quietly into the Welland Canal which carries her over 30 miles, until she passes out again upon the shallowest of the great lakes, Lake Erie. Onwards for another 140 miles, and then through similar works she reaches Lake Huron. Through a wonderful archipelago of islands, scattered on the water on its northern shore, she wends her way until the old French post called the Rapids of St. Mary is seen upon the low and wooded shores. Here for the first time in her long inland voyage she has to leave Canadian territory, the canal which takes her onwards being built on American ground [1]. And now at last she will have arrived at the ultimate stage of her wanderings, for before her stretch the 400 miles of the deeps of Lake Superior, 600 feet above the level of the sea.'

(7) **The Lakes.**—Besides Lakes Ontario, Erie, Huron, and Superior, many lesser lakes may be noticed in

---

[1] A new canal is being constructed in Canadian territory.

Ontario. There is *Lake St. Clair*, placed as a link between Lakes Erie and Huron.

Lake Simcoe, the scene of early missionary efforts amongst the Hurons.

Lake Nipissing, on the old canoe route to the West, now on the track of the railway which points westward to Sault Ste. Marie.

Lake Couchiching, the 'Lake of many Winds.'

Lake Muskoka, described as a lovely sheet of water dotted with picturesque islands with steep winding banks.

Lake Rosseau, a lake of irregular shape and filled with picturesque islands and fringed with noble forests.

The greater Georgian Bay, an eastern arm of Lake Huron, may be regarded almost as a lake in itself.

Lake Nipigon, called by some 'the most beautiful of all lakes.'

Lake of the Woods and *Rainy Lake*, which feed first of all the great St. Lawrence River.

(8) **Divisions**.—Ontario is divided into 46 counties, which may be grouped as follows:—

I. Five in Lake Erie District, viz. Essex, Kent, Elgin, Norfolk, Haldimand.

II. Three in the Niagara District, viz. Monck, Welland, Lincoln.

III. Three in Lake Huron District, viz. Lambton, Huron, Bruce.

IV. Seven in the Inland District, viz. Middlesex, Perth, Oxford, Waterloo, Wellington, Cardwell, Brant.

V. Two in the Georgian Bay District, viz. Grey and Simcoe.

VI. Three in the Northern District, viz. Victoria, Haliburton, Peterborough.

VII. Seven in the Lake Ontario District, viz. Wentworth, Hatton, Peel, York, Ontario, Durham, Northumberland.

VIII. Four in the Quinté District, viz. Hastings, Lennox, Addington, Prince Edward.

IX. Six in the St. Lawrence River District, viz. Frontenac, Leeds, Grenville, Dundas, Stormont, Glengarry.

X. Five in the Ottawa River District; viz. Renfrew, Lanark, Carleton, Russell, Prescott.

In addition to these there is the great Manitoulin Island on the north of Lake Huron, which is rapidly being settled.

In the above counties there are 88 electoral districts. The power of self-government is carried out in the Dominion. Any village with 750 inhabitants may be incorporated under the Municipal Acts. If such a village increases to 2000 it acquires the status of a town, and if it reaches 15,000 it gains the full dignity of a city.

(9) The shipping of Ontario is on a great and increasing scale. The chief ports are as follows :—

Names.	County.	Lake or River.
1. Ottawa	Carleton	on the Ottawa.
2. Brockville	Leeds	
Cornwall	Stormont	on the St. Lawrence.
Kingston	Frontenac	
Morrisburg	Dundas	
3. Belleville	Hastings	
Cobourg	Northumberland	
Hamilton	Wentworth	
Napanu	Lennox	
Oakville	Halton	on Lake Ontario.
Port Colborne	Northumberland	
Port Hope	Durham	
Pictou	Prince Edward	
Toronto	York	
Whitby	Ontario	
4. Niagara	Lincoln	on the Niagara.
St. Catherine's	Welland	
5. Amherstburg	Essex	
Dunville	Monck	
Port Burwell	Elgin	on Lake Erie.
Port Dover	Norfolk	
Port Rowan	Norfolk	
Port Stanley	Elgin	

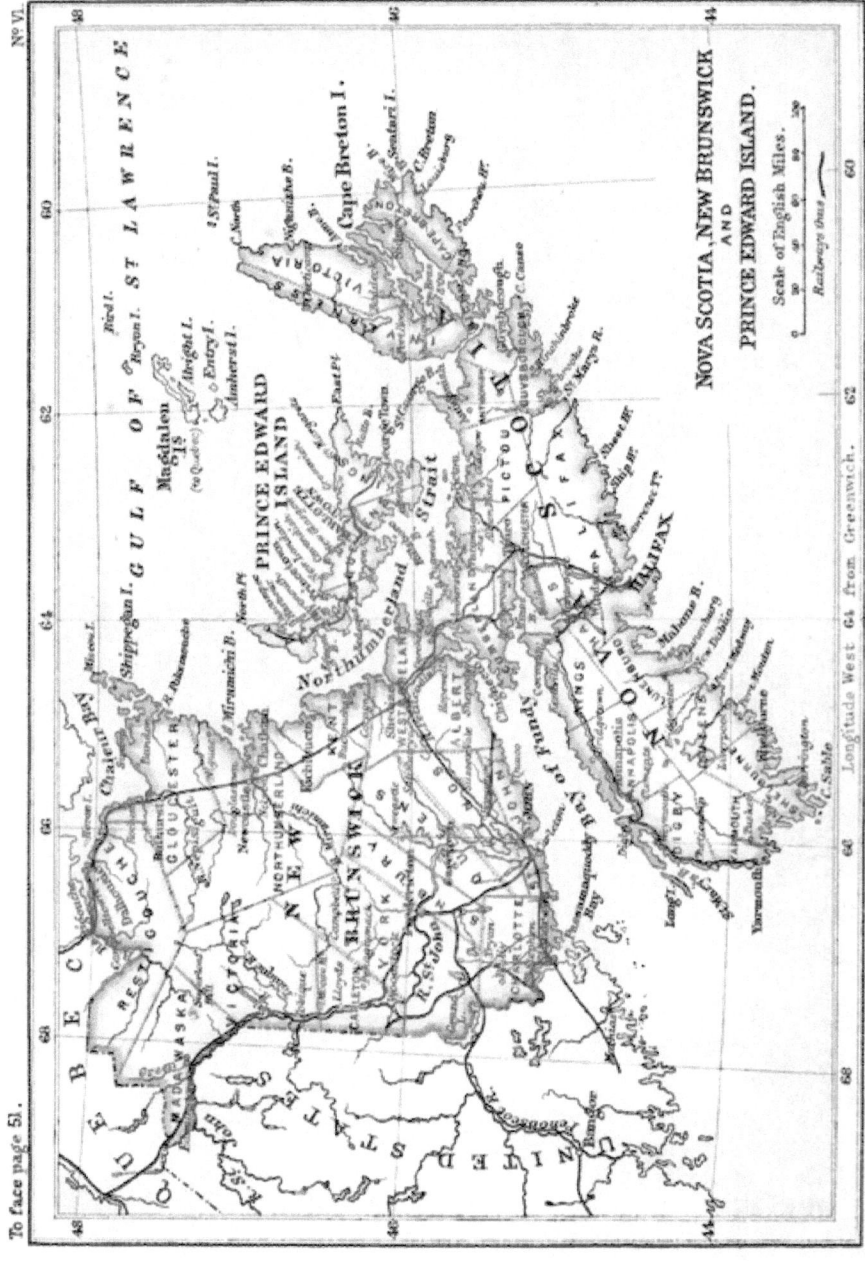

Names.	County.	Lake or River.
6. Chatham	Kent	} on Lake St. Clair.
Windsor	Essex	
7. Goodrich	Huron	on Lake Huron.
8. Collingwood	Simcoe	} on Georgian Bay.
Sydenham	Grey	
9. Sault Ste. Marie		on Lake Superior.

Of these the most important are—

		Arrivals and Departures.
Kingston,	with a tonnage of	1,066,592
Toronto,	,, ,,	906,704
Ottawa,	,, ,,	727,796
St. Catherine's,	,, ,,	413,231
Hamilton,	,, ,,	129,175

including coasting trade [1].

CHAPTER IV.

Nova Scotia.

(1) OF all the Provinces of the Canadian Dominion, Nova Scotia has the most interesting and romantic history. Around its stormy coasts and bleak capes, the great struggle between the French and British was being incessantly carried on in the seventeenth and eighteenth centuries. The ruins of Louisburg, on the north of Gabarus Bay, are memorials of the broken Power which aimed at exclusive dominion in North America, from the St. Lawrence to the Lakes and Western

[1] See General Map No. 1, with list of Shipping and Population.

Prairies. By the twelfth article of the Treaty of Utrecht, 'All Nova Scotia or Acadia comprehended within its ancient boundaries, as also the city of Port Royal, now called Annapolis, were yielded and made over to the Queen (Anne) of Great Britain and her Crown for ever.' Much of the romance of early colonisation has taken place in Nova Scotia. Those 150 adventurers styled The Knights Baronets of Nova Scotia, created by Charles I, give us a specimen of that determined spirit of colonisation of the New World in the wake of the Elizabethan heroes, which was characteristic of Englishmen. Little by little the British advanced, and fought many a sanguinary battle for the possession of Port Royal, Canso, and Cape Breton, the key of the St. Lawrence in those days. At last the struggle closed with the expulsion of the Acadians from their country. The poet Longfellow has, with deep pathos, made the village of Grand Pré, Blomidon and the Basin of Mines historical.

Nova Scotia, as the principal Maritime Province, has its peculiar and distinct features. The colonists who live there are chiefly fishermen and toilers of the sea, like those of Newfoundland. But mining and lumbering are carried on also to a large extent, and the coal trade is steadily growing. 1,576,692 tons being sold in 1888, while very much less than half was sold ten years ago[1].

(2) **Boundaries.**—The Province of Nova Scotia is situated between 43° and 47° N. latitude and 60° and 67° W. longitude, comprising the Peninsula of Nova Scotia and the Island of Cape Breton. It is bounded on the north by Northumberland Strait and the Gulf of St. Lawrence, on the east and south by the Atlantic Ocean, on the north-west by the Bay of Fundy and the Bay of Chignecto. It is long and narrow in shape, running in a north-easterly and south-westerly direction.

[1] See p. 48 of 'Official Handbook,' 1890.

Area.—Nova Scotia is about 300 miles long and 100 miles broad. Its coast-line is 1200 miles, and the whole area is 20,000 square miles, of which one-fifth is covered with lakes and small rivers. It is about one-ninth the size of Ontario.

(3) **Surface.**—There are no very high mountains in Nova Scotia. The central watershed extends down the whole length, giving a northerly and southerly slope. The highest mountains are found along Cape Breton in latitude 46° N. The South Mountains are in Annapolis and King's counties, and form the western part of the central ridge. The North Mountains run parallel with them and border on the Bay of Fundy. Between them flows the Annapolis River, along a most sheltered and fertile valley, which was the home of the Acadian peasantry, celebrated by the poet Longfellow in 'Evangeline.' The Cobequid Mountains lie in the county of Colchester, north of the Basin of Mines. The boldest scenery is found in Cape Breton from Inganische to St. Anne's Bay.

(4) **Coast.**—Nova Scotia and Cape Breton, which are divided by the narrow Gut of Canso, have a very long coast-line, deeply indented with innumerable bays, especially on the Atlantic sea-board. It is, therefore, especially adapted for a seafaring and maritime population. The Bay of Fundy is remarkable for its tides, especially on its eastern arm, which terminates in the Bay of Mines. At the Equinoxes the tides rise sometimes to the height of fifty feet, and form a bore similar to that seen in the Bristol Channel.

Rivers.—There are no long rivers in the peninsula, owing to its natural configuration. No river exceeds fifty miles in length. The largest are the *Shubenacadie*, flowing into the east corner of Mines Basin, *East River of Pictou* into Pictou Harbour, *St. Mary's La Have, Mersey* into the Atlantic on the south, *Annapolis* into Annapolis Basin, and *Cornwallis* into Mines Basin.

Lakes.—There are many lakes, but none exceed twelve miles in length. The chief are: *Darkies, Rossignol, Malaga* in the south; *Ship Harbour Lake* and *Lake Ainslie* in Cape Breton. *Bras d'Or* is the great lake in the interior of Cape Breton.

(5) **Islands.**—*Isle Madame*, near the Gut of Canso, is sixteen miles long, and has a large fishing population. *Boularderie* is a long-shaped island north of Cape Breton, and forms part of Victoria County in Cape Breton. *Sable Island* lies 100 miles south of Cape Breton where La Roche's convict colony was starved in 1598, and *St. Paul's Island* lies beyond Cape North. Amongst others are *Pictou, Scatari, Cariboo, Tancock, Long Island, Briar Island.*

Capes.—The principal capes are: *Chignecto, Split, D'Or, Blomidon, Mulagash, St. Lawrence, St. George, Egmont, Granby, Dauphine, Sambro Head, Breton, Sable.*

Bays.—On the Bay of Fundy: *St. Mary's Bay, Grand Passage, Digby Gut, Annapolis Basin, Mines Basin, Cobequid Bay, Chignecto Bay, Cumberland Basin.*

On the Northumberland Strait: *Baie Verte, Pugwash Harbour, Pictou Harbour, George Bay.*

On the Atlantic side: *Chedabucto Bay, Milford Haven, Torbay, Sheet Harbour, Musquodoboit Harbour, Halifax Harbour, Bedford Basin, Lunenburg Harbour.*

Nova Scotia possesses unrivalled opportunities for fishing, and the annual value of her fish exceeds $1\frac{1}{2}$ millions sterling, a large portion of which is exported. She also exports enormous quantities of tinned lobsters to all parts of the world. Her best markets are the West Indies, and then the United States.

(6) **Climate and Soil.**—The climate of Nova Scotia, especially on the south and western portions, where it is

sheltered by the high ground, is far milder than that of the inland provinces. The Gulf Stream sweeps up from the Gulf of Mexico within a short distance of its southern shores. The average temperature of Annapolis County is 8° warmer than that of Cape Breton, which is a bleak and stormy region, and 6° warmer than that of the State of Massachusetts. The Valley of Annapolis is famed for its beautiful orchards, and fruits of many descriptions.

Although the peninsula is not an agricultural country in any remarkable degree, its chief productions being those of the sea, the mine, and the forest, still there are some very rich tracts, especially in Hants, King's, and Annapolis counties[1]. The average production of the crops of Nova Scotia are as follows:—

Wheat	yield per acre	18 bushels
Rye	,, ,,	21 ,,
Barley	,, ,,	35 ,,
Oats	,, ,,	34 ,,
Buckwheat	,, ,,	33 ,,
Indian Corn (Maize)	,, ,,	42 ,,
Turnips	,, ,,	420 ,,
Potatoes	,, ,,	250 ,,
Mangold Wurtzel	,, ,,	500 ,,
Beans	,, ,,	22 ,,
Hay	,, ,,	2 tons.

The ungranted lands in the different counties amounted in 1887 to 2,116,811 acres, of which 236,000 were in Halifax County, 475,000 in Victoria County, 202,000 in Yarmouth County, 139,000 in Annapolis County, 136,000 in Shelburne County, 135,000 in Queen's County, 147,000 in Inverness County, 114,000 in Lunenburg County, 101,000 in Guysborough County[2].

But it is impossible for Nova Scotia to compete with

---

[1] Appendix V.
[2] See p. 53 of 'Official Handbook,' 1888.

the great grain-producing region of the North-West. In the future she will probably become a great manufacturing centre. Her coal-fields at Pictou County, Cumberland County, and Cape Breton, combined with her iron, gold, and other mineral wealth, must give her a leading position. The Marquis of Lorne has written :—' If wages were only as low in Nova Scotia as they are in England and Scotland, one of her ports—the port of Pictou—would soon rival Glasgow or Belfast or London as a great iron ship-building port. Near it are mines almost as vast as those of Lanarkshire, close to the water are great veins of coal of twenty to thirty feet in thickness.'

Her geographical position also is very favourable for the distribution of her manufactures to all parts, whether by sea or land. The railway across the isthmus connects her with the whole Canadian system and the Pacific coast. From Halifax to Vancouver, a distance of about 3000 miles, there is uninterrupted communication.

(7) The following is the list of the counties and chief towns of Nova Scotia and Cape Breton :—

Counties.	Chief Towns.
Annapolis	Annapolis.
Antigonish	Antigonish.
Colchester	Truro.
Cumberland	Amherst.
Guysborough	Guysborough.
Halifax	Halifax.
Hants	Windsor.
King's	Kentville.
Lunenburg	Lunenburg.
Pictou	Pictou.
Queen's	Liverpool.
Shelburne	Shelburne.
Yarmouth	Yarmouth.

CAPE BRETON.

COUNTIES.	CHIEF TOWNS.
Cape Breton	Sydney.
Inverness	Port Hood.
Richmond	Arichat.
Victoria	Baddeck.

(8) Halifax, the capital of the Province, has a population of 40,000[1], and occupies the west side of what was once called Chebucto Bay. It was called after Lord Halifax, the President of the Board of Trade and Plantations (1749).

It is one of the two British combined naval and military stations in British North America, and as an Imperial stronghold is of the greatest value. It lies on the highway to the East and the Pacific, being connected with British Columbia by the Canadian Pacific Railway. It possesses one of the finest harbours in the world. Sir Charles Lyell, in an account of his travels in North America, remarks that the harbour of Halifax reminded him more of a Norwegian fiord such as that of Christiania than any other place he had seen.

(9) **Population.**—The population of Nova Scotia is about 500,000, well distributed over the counties; nearly three-fourths of them being Protestants, and the remainder Roman Catholics. Their chief industry is in their 'fisheries,' and Nova Scotians own more shipping in proportion to their numbers than any other country in the world. Along the deeply indented coast-line there is an admirable nursery for seamen. If there were a North Atlantic squadron of a combined Colonial and Imperial fleet, Nova Scotia would furnish a large contingent. The number of vessels belonging to the port of Halifax alone exceeds 1000. Yarmouth owns a smaller number, but the tonnage is greater.

---

[1] See p. 49 of 'Official Handbook,' 1890.

In addition to the chief towns already given, may be mentioned Barrington, Digby, Londonderry, Maitland, Pugwash, Parrsborough, Port Hawkesbury, Port Medway, Weymouth. Many names and towns in the Province remind us of historical events in and near the Peninsula. Rossignol, now Liverpool, on the south-east coast, was a name given by De Monts, 1604, to a place where he found a fellow-countryman trespassing on his patent. Lunenburg close by indicates the German immigration in 1753, Liverpool on the Mersey marks on the coast itself the English occupation. Torbay, Dartmouth, and Bridgwater recall the west-country names and probably indicate a local migration. Cornwallis, named after the ill-starred soldier, recalls less agreeable memories of surrender and defeat in the adjoining continent; in Shubenacadie the old French name for the Peninsula has been preserved, and perhaps Canso, Chebucto, Chedabucto, Cobequid, Chignecto recall the earlier nomenclature of the aboriginal Micmacs or perhaps of the Malicites, tribes of the Algonkin family.

CHAPTER V.

New Brunswick.

(1) **Boundaries.**—The Province of New Brunswick is situated between the Restigouche River and Bay Chaleur on the north, and the Bay of Fundy on the south; between the Gulf of St. Lawrence and Northumberland Strait on the east, and the United States on the West.

Area.—Its area is 27,322 square miles, equal to 17,486,280 acres. It is 210 miles in length and 180 miles broad.

Surface.—New Brunswick is connected with Nova Scotia by the narrow isthmus of Chignecto. It has a coast-line of 500 miles, being deeply indented with bays, harbours, and inlets. On the north-east side there are very few hills, and the most picturesque side is towards the Province of Quebec, where the hills rise to a height of 500 to 800 feet, covered with lofty forests nearly to their summits.

(2) **Rivers.**—*Restigouche*, an Indian name, meaning 'the broad,' or, according to others, the 'five-fingered' river. It is 200 miles long.

The *Nepisiguit* is eighty miles long and flows very swiftly. It empties into Bathurst Bay, an arm of Chaleur Bay.

The *Miramichi* flows into Miramichi Bay on the east coast, and is navigable for some distance.

The *Peticodiac* flows into Shepody Bay on the south.

The *St. John* is the largest river in New Brunswick, and is 450 miles long, being navigable for large steamers as far as Fredericton, a distance of eighty-five miles, and for small steamers as far as Woodstock, a distance of 150 miles. The river is noted for its beautiful scenery and for its magnificent 'Grand Falls,' seventy-four feet high. Amongst its tributaries are the Oromocto, Madawaska, Tobique, the Washademoak and the Kennebecasis in the Province itself, and Aroostook and Allagash from the State of Maine. The St. John rises in Lake Temiscouata, close to the south bank of the St. Lawrence.

Amongst the smaller rivers on the east coast are the Richibucto and Cocagne. The St. Croix is on the United States frontier and flows into the Bay of Fundy.

Everywhere in this Province we shall find rivers and

streams affording easy access from one place to another. They abound in every kind of fish useful for the purposes of man. The value of the fish taken annually is calculated to be £500,000. The River Restigouche is considered to be one of the finest salmon-fishing rivers in the world, although perhaps not equal in this respect to the Cascapedia, further north in the Province of Quebec.

Lakes.—Although New Brunswick cannot boast of such lakes and rivers as her sister provinces to the west, she possesses, nevertheless, some very fine sheets of water.

Grand Lake in Queen's County, twenty-eight miles long, with a width of six miles, and communicating with St. John's River.

Washademoak, also in the Queen's County.

Maquapit and *French Lake*.

Temiscouata Lake at the head of the Madawaska River.

Loon Lake, *Eel Lake*, and the *Oromocto*, form a chain along the boundary in the county of York.

The *Miramichi*, *Salmon*, *Nepisiquit* and *Nictaux* are in the eastern division of the Province.

(3) **Coast.**—The Province of New Brunswick has a very long coast-line, with many fine harbours on all sides. Chaleur Bay extends along the north for about ninety miles, and affords access far into the interior. It includes two minor bays, Nepisiquit and Caraquet Bay.

Miscou and Shippegan are two islands on its southern extremity.

Miramichi Bay is on the Gulf of St. Lawrence, and facing Northumberland Strait are Shediac Bay and Baie Verte.

On the south is the Bay of Fundy, noted for its strong currents and tides. Amongst its bays are Cumberland Basin and Shepody Bay, St. John Harbour, and Passamaquoddy Bay on the extreme south-west corner.

(4) **Population.**—The population of New Brunswick was in 1881 computed at 321,000. Most of them are of British and Irish descent, but there are many descendants of the early French settlers.

The various Protestant bodies include about two-thirds of the population, the Roman Catholics one-third. The capital, Fredericton, has a Cathedral of the Episcopal Church, and is the seat of the present Metropolitan of Canada. The date of the founding of the Bishopric is 1845.

The Indians number a little over 1000. Their reserves are found on the Tobique River in Victoria County, and on the St. John and Madawaska Rivers. The Micmac Indians, whom we hear of in the early history of the Colony, number 913, and the Malicites 500.

(5) **Divisions and Counties.**

St. Lawrence Counties.	*County town.*
Restigouche	Dalhousie.
Gloucester	Bathurst.
Northumberland	Newcastle.
Kent	Richibucto.
Westmoreland	Dorchester.
Bay of Fundy.	
Albert	Hopewell.
St. John	St. John.
Charlotte	St. Andrew's.
Inland Counties.	
King's	Hampton.
Queen's	Gagetown.
Sunbury	Oromocto.
York	Fredericton (capital).
Carleton	Woodstock.
Victoria	Grand Falls or Colebrooke.

(6) **Towns.**—*Fredericton* is the capital, with a population of 7,000. The largest city and most important port is St. John, with a population of 50,000, including Portland. The tonnage of its shipping is very large, the timber from the State of Maine being shipped through its port. This city has been called 'the Liverpool of British America.'

Following the coast-line east and north the following are the most important ports. Hopewell and Dorchester at the head of the Bay of Fundy, Shediac, Chatham, Bathurst, Dalhousie, Campbelltown along Northumberland Straits and the Gulf of St. Lawrence. Other rising towns are Moncton (6,000), Sackville, St. Stephen, Harvey and Bay Verte.

The fishing industry is very important in New Brunswick, and in 1888 the number of boats and vessels was calculated at 4863, employing 9840 men[1].

(7) The chief wealth of New Brunswick lies in its forests, as the following table of exports will show[2]:—

	£
Products of the Mines	19,000
Fisheries	157,000
Products of the Forests	736,000
Animals and their produce	80,000
Agricultural	50,000
Manufactures	75,000
Miscellaneous	10,000

(8) **Railways.**—The chief railway in New Brunswick is 'The Intercolonial,' the first section of which was opened between St. John and Shediac in 1860. It was finished in its entire length, between Halifax and the St. Lawrence, in 1876. This railway enters New Brunswick at Dorchester, and skirts the eastern coast to Chaleur Bay and Campbellton. Thence it enters into

---

[1] See p. 53 of 'Official Handbook,' 1890.
[2] See 'Official Handbook,' 1890.

Bonaventure and the Province of Quebec until it reaches Riviere du Loup on the St. Lawrence. On the south of New Brunswick there is a branch connecting with St. John and Fredericton, and with the United States Railway system to the West. By means of this communication New Brunswick is on the high road to Quebec and the Far West in one direction, and to the State of Maine and the United States in another. The through distance from Halifax to Riviere du Loup and Quebec is 561 miles.

This intercolonial railway, therefore, links the Naval Station of Halifax with the old fortress of Quebec. Those travellers who wish to reach the Far West may either take their passage by steamer straight to Quebec or to Halifax. It may be noticed that the Canadian Pacific Railway Company has a line to St. John from Montreal, running by way of Sherbrooke through the State of Maine in a south-easterly direction. This is the most direct of all routes, and saves several hours in reaching the water on the Atlantic coast.

Those who wish for the shortest route between Europe and America, advocate a line of communication from Newfoundland at some point on St. George's Bay, to Shippegan at the corner of Chaleur Bay in New Brunswick. In this case, Newfoundland and then New Brunswick will be on the main line of communication. The distance by sea between the nearest point of the Irish coast and Newfoundland is only 1640 miles, occupying less than four days' steaming, then comes the railway journey across Newfoundland to St. George's Bay, and a short run to New Brunswick, thus linking with the present system at Bathurst.

(9) **Immigration.**—The following account has been given of the immigration into New Brunswick:—'The position of New Brunswick is not favourable to immigration. The stream of immigrants from Europe divides into

two main portions, passing respectively north and south of the Province. The great river St. Lawrence, with the wealthy and populous towns on its banks, and the great chain of Canadian Lakes, with the rich north-west prairie lands, attract emigrants to the north, while the pushing industries and untiring enterprise of the United States draw off others to the south, so that the really valuable lands in New Brunswick have been too much overlooked. However, according to official returns (1890), there are now about fifty free grant settlements in the Province, settled by thousands of industrious men who had no means of purchasing farms.

The available lands in the various counties of New Brunswick amount to 6,000,000 acres, and are classified as 'upland,' 'intervale¹' and 'swamp.' In the future New Brunswick may become a farming country, although at present her main wealth consists of her magnificent forests and fisheries.

Out of the total acreage of the Province of New Brunswick (17,300,000 acres) about 10,000,000 have been granted and located and 7,400,000 are still vacant. Of this 1,900,000 are in the County of Northumberland, 1,600,000 in Restigouche, 820,000 in York, 630,000 in Gloucester, and 550,000 in Kent. The proximity of extensive coal fields must help to develop the manufacturing industries of the country. At present there are four large cotton-mills in the Province giving employment to about 1300 people².

Prince Edward Island.

(10) This island is the most thickly populated Province of the Dominion, and is traversed by a railway from end

[1] Appendix VI.
[2] Extract from 'Official Handbook,' issued by the Canadian Government, October, 1888.

to end, from Georgetown to Tignish. Its area is 2,135 square miles or 1,365,400 acres, and the population is about 109,000. It lies between 46° and 47° N. latitude and between 62° and 64° W. longitude. Its length is about 140 miles and its greatest breadth thirty-four miles. It lies in the Gulf of St. Lawrence, and is distant nine miles from New Brunswick, fifteen miles from the mainland of Nova Scotia, and thirty miles from Cape Breton Island.

Climate.—Owing to its insular position, Prince Edward Island enjoys a more equable climate than the inland Provinces. The population, unlike that of Nova Scotia and New Brunswick, is mainly agricultural, and the most important products are oats, barley and potatoes. A larger proportion of land is under cultivation here than in any other province. Much of the produce is exported. Ship-building is also carried on extensively, and the fisheries round the island are very productive. A great trade in canned lobsters is being carried on. The principal coast waters are Richmond Bay, Cardigan Bay, Hillsborough Bay, Bedeque Harbour and Egmont Bay. The surface of the island is low and undulating, and the watershed runs east and west, so all the rivers are small. The soil of the island is very fertile, being composed largely of the 'alluvium' or deposit of the St. Lawrence river[1]. For three months of the year[2]—from December to March—communication with the island is rendered difficult by the ice round its coasts. It was called Prince Edward Island after Prince Edward, Duke of Kent, and came first into the possession of the British by the Treaty of Paris. Its first name was St. John's Island.

(11) Early Settlements.—This island was first colonised in 1715, but immigration was slow, there being in 1752

[1] Appendix VII.
[2] See p. 12 of 'Agricultural Canada,' 1889.

only 1354 inhabitants. The first Governor was appointed in 1780, the whole population not being more than 5000. In 1803, Earl Selkirk, who colonised the North-West Territories, settled about 800 Highlanders here.

Counties.—Prince Edward Island is divided into three counties, *Prince*, *Queen's*, and *King's*. The first named lies to the north, *Queen's* in the centre, and *King's* on the south.

(12) **Towns.**—The chief towns are *Charlottetown*, the capital, with a population of 12,000, lying about the centre of the south coast of the island, with a fine harbour.

Georgetown lies about thirty miles east of Charlottetown. It is the chief port of the island, and has a magnificent harbour, which remains open nearly all the year.

Summerside is the second town in trade and population, and lies on Bedeque Bay.

Tignish is at the extreme north-west end of the island, fifty-five miles from Summerside, where a large fleet of fishing boats is employed in catching mackerel.

Souris is at the eastern corner of the island. At *Alberton*, also, there is a busy trade carried on by a fishing population.

There is quick and easy communication by boat and rail between all parts of this prosperous little island. It has 210 miles of railway in operation. The distance between Bedeque Bay and Shediac Bay in New Brunswick is short, and communication is gained there with the Canadian Railway system.

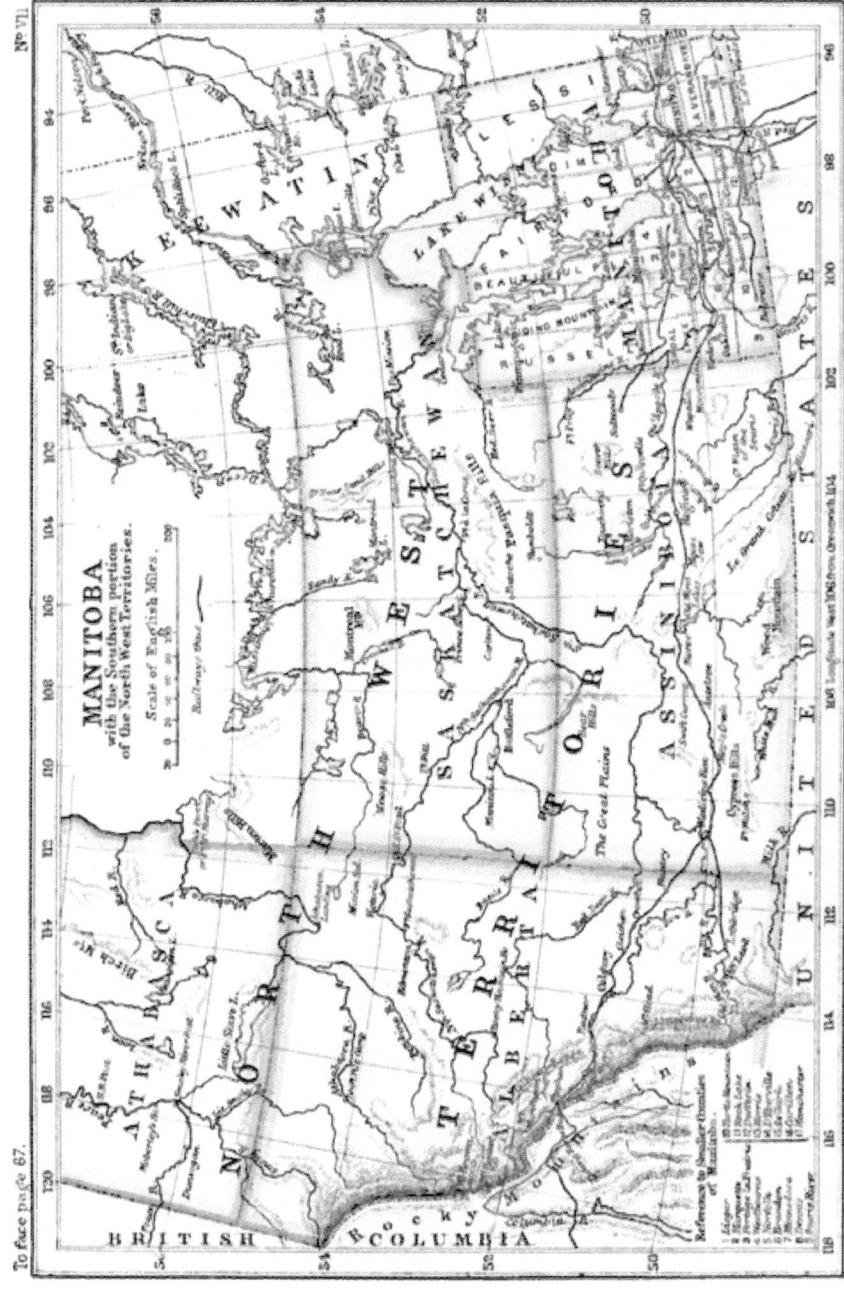

CHAPTER VI.

Manitoba.

(1) MANITOBA is a province carved out of the Domain over which the Hudson's Bay Company used to exercise their rights, the rest of the country being now called the North-West Territories. It consists almost wholly of prairie land, and the settler has no trouble to begin with in tree-felling and log-burning in order to clear the soil. In Manitoba are wide, open, and limitless plains, and, in contradistinction to the River and Lake Provinces of Quebec and Ontario, it is called 'the Prairie Province.' Before the Province was settled it was known as part of the Hudson's Bay Territory, the Selkirk Settlement, Red River Country, and Rupert's Land. As a home for settlers its history dates from 1811–16, when Thomas Douglas, Lord Selkirk, brought a number of Scotchmen out under his own care and supervision. He reached the Red River not by the St. Lawrence, but by Hudson's Bay, during the few summer months when navigation is possible. For a long time the Red River Settlement was an isolated tract of country, the nearest point to the head of Lake Superior being Port Arthur, distant 435 miles.

The word Manitoba is a contraction of two Indian words 'manitou' (spirit), and 'waban' (strait), and was originally applied to a lake which seemed to the imaginative Indians to be troubled by some god or spirit. It is pronounced Manitōba.

(2) **Boundaries.**—On the east and north lie Ontario

and the Keewatin district; on the south the International Boundary Line, on the west the districts of Assiniboia and Saskatchewan. It occupies a central position in the continent, being half-way between the Pacific and Atlantic Oceans. It lies between 49–53° N. latitude, and 90–101° W. longitude. Its shape is that of a parallelogram.

Area.—Its area is calculated to be 60,520 square miles, or as nearly as possible one-half the size of Great Britain and Ireland. The slope of the country is towards the north. The Riding Mountain and the Duck Mountain to the north, rising nearly 2000 feet, and the Turtle Mountain to the south on the Boundary Line are the only elevated points in this province, its surface being for the most part level or undulating. In the words of the Earl of Dufferin (1877), 'From its geographical position and its peculiar characteristics Manitoba may be regarded as the keystone of that mighty arch of sister provinces which span the continent from the Atlantic to the Pacific. It is here that Canada, emerging from her woods and forests, first gazed upon her rolling prairies and unexplored North-West, and it was hence that as a prelude to future expansion she took a fresh departure.' Winnipeg is 1423 miles west of Montreal, about 200 miles further than Madeira from London, and the creation of the Province of Manitoba at so great a distance from the old province is another illustration of the immensity of our North American domain. Yet, Manitoba is only in the centre of the continent after all, and is equidistant from British Columbia on the west and Quebec on the east. The prairies seem absolutely limitless, and they stretch to the base of the 'Rockies,' which by their trend to the north-west make the central plateaux of British North America wider than in the more southern latitudes of the United States[1]. This immense region, for

[1] Appendix VIII.

hundreds of miles, is famed for its grains and grasses. Acre for acre, the country here could support a thicker population than any similar tract on the globe. It is on the wonderful capacity of Manitoba and the North-West for producing grain and vegetables that the assumption is based that British North America could sustain a population of nearly 200,000,000; the northern limit for wheat being placed by Hurlbert at 58° N. lat.

(3) **Population.**—The Province of Manitoba is in its infancy. In 1870 the population was 11,965, of whom only 1614 were whites. Since then the latter have rapidly increased, and were estimated in 1887 at over 112,000, of whom about three-fourths are British. The city of Winnipeg alone is calculated to number 22,000 [1].

Besides British there are a large number of Germans, French and English half-breeds, and Icelanders. Of all nationalities the Scotch have been perhaps the most persistent and successful colonists of British North America. The Mennonites or German makers form a peculiar element in the population of Manitoba. They are settled along both banks of the Red River in twenty-five districts or reserves, embracing 512,000 acres, of which 300,000 are under settlement.

(4) **Lakes.**—Manitoba, and the country to the north and north-west, is covered with a network of rivers and lakes extending far north. It is calculated that the great Lake Winnipeg drains a basin of 400,000 square miles, an area of country almost equal to the St. Lawrence Valley. Unlike Lake Superior and the others to the west, Lake Winnipeg is comparatively shallow, and is nowhere more than 70 feet deep. Its elevation above the sea is 710 feet. Its length is 280 miles, and its breadth 57 miles [2].

[1] See p. 68 of 'Official Handbook,' 1890.
[2] See p. 416, Hayden and Selwyn.

Lake Winnipegosis, or Little Winnipeg, is 120 miles long, 27 miles broad, and covers 2000 square miles. This lake pours its superfluous waters through the Water-Hen River into *Lake Manitoba*, which is 120 miles long, about 24 miles broad, and covers 1900 square miles. This lake, in its turn, pours its waters into Lake Winnipeg.

(5) The **Rivers** of this region are:—

Winnipeg River, 163 miles long, flowing north-west from the Lake of the Woods, and gathering in its course the waters of many small lakes.

The *Red River*, nearly 600 miles in length, flowing from the south, and entering Lake Winnipeg.

The *Assiniboine*, from the north-west, joining the Red River at Winnipeg, and carrying the waters of the Souris from the south.

The *Barens River*, draining some lakes from the east.

Through the *Little Saskatchewan* the accumulated waters of Lakes Winnipegosis and Manitoba flow into Lake Winnipeg, the communication between the two smaller lakes being by the Water-Hen River.

The waters of Lake Winnipeg, which are gathered from so many different sources as far asunder as the Lake of the Woods through the River Winnipeg on the east, the 'Rockies' through the Great Saskatchewan on the west, and the hills of Minnesota through the Souris on the south, flow finally into Hudson's Bay on the north by the great Nelson River. This river is 350 miles long, and passes through a large number of small lakes. Its course is rendered difficult to navigate by the countless numbers of rapids. Its volume of water is enormous. Communication with the north-west, *viâ* Hudson's Bay and the Nelson River, is carried on with great difficulty, the Hayes route from York Factory being the usual inland route from Hudson's Bay. In the first place, the great bay itself is frozen over for six months in the year; and in the second place, there is no navigable river which can take the colonist and explorer into the heart of the continent from the bay during the brief summer months. In Manitoba itself we have the key to understanding the drainage and slope of the country surrounding Lake Winnipeg, for on the south the Winnipeg River opens up the route eastwards by lake and river to Thunder Bay, whilst on the north the Great Saskatchewan is 'the high-road to the north-west.' In course of time a railway may link Winnipeg with Hudson's Bay.

(6) **Climate.**—There is no very marked difference between the climate of Manitoba and that of Quebec and Ontario. It is warmer in summer and colder in winter. Spring opens up at nearly the same time along the country from Lake Superior to Athabasca:—

'The juncture of the seasons is not very noticeable. Spring glides insensibly into summer, summer into fine autumn weather, which, during the equinox, breaks up in a series of heavy gales of wind accompanied by rain and snow. These are followed by that divine aftermath, the Indian summer, which attains its true glory only in the north-west. The haziness and dreamy fervour of this mysterious season have often been attributed to the prairie fires, which rage over half a continent in the fall, and evoke an enormous amount of heat and smoke.' The winters are also described:—

'The winters of the north-west, upon the whole, are agreeable and singularly steady. The mocassin is dry and comfortable throughout, and no thaw, strictly speaking, takes place till the spring, no matter how mild the weather may be. The snow, though shallow, wears well; and differs greatly from eastern snow. Its flake is hard and dry, and its gritty consistence resembles white slippery sand more than anything else. Generally speaking, the further west the shallower the snow, and the rule obtains even into the heart of the "Rocky Mountains." In south-western Ontario the winter is milder, no doubt, than at Red River, but the soil beats the soil of Ontario out of comparison; and, after all, who would care to exchange the crisp, sparkling, exhilarating winter of Manitoba for the rawness of Essex in south Ontario?"[1]

The mean winter temperature of Manitoba is 5° below freezing point, and for the summer 65°. In the depth of winter the atmosphere is generally calm and still, and the cold is, therefore, not felt in the proportion we should imagine in the raw, damp, and blustering climate of the British Islands. Probably there is no weather in North America so disagreeable as the cold raw spring-weather of England when the winds blow from the east.

(7) **Divisions.**—Manitoba is divided into four counties: Selkirk, Provencher, Lisgar, Marquette. These are subdivided into twenty-four districts or electoral divisions, returning 38 members to the provincial assembly:—

Westbourne.	St. Paul's.
Portage la Prairie.	St. Andrew's, S.
Poplar Point. }	St. Andrew's, N.
High Bluff. }	St. Clement's.
Baie St. Paul.	Rock Wood.
St. François Xavier, W.	Springfield.
St. François Xavier, E.	St. Boniface.
Headingley.	St. Boniface.
St. Charles.	St. Vital.
St. James.	St. Narbert.
Winnipeg.	St. Agathe.
Kildonan.	St. Anne.

[1] Extract from Bryce's 'Manitoba.'

(8) **Towns.**—*Winnipeg.* In 1878 the first railway ran into Winnipeg, and since then a new era began for the North-West, and the prosperity of the town commenced. Standing at the confluence of the Red River and Assiniboine it is the doorway to the great prairie region beyond, and is a great distributing centre. In 1874 the assessment of real and personal property was $2\frac{1}{2}$ million dollars; in 1883 the assessment of the town had risen to $32\frac{1}{2}$ million dollars. The site of Winnipeg was occupied originally by a post of the Hudson's Bay Company, and was known as 'Upper Fort Garry.'

Portage la Prairie[1] (2500) is 56 miles west of Winnipeg along the railway, and is becoming a mercantile and manufacturing centre. *Brandon* is a rapidly rising town further west, with a population of 3000. Amongst others there is Selkirk (1000), Emerson (800), and other smaller towns and villages. Gladstone, Minnedosa, Shoal Lake, and Birtle are on the Manitoba and North-West Railway, which is opening up the Great Saskatchewan valleys.

CHAPTER VII.

North-West Territories.

(1) UNDER this general term may be included all the land of the Canadian Dominion not contained in Quebec, Ontario, Nova Scotia, Prince Edward Island, New Brunswick, the North-East Territories, Manitoba and British

---

[1] See p. 69 of 'Official Handbook,' 1890.

Columbia. These territories[1] are the great unorganised domain of North America, and stretch over a vast tract of country far northward to the Arctic Ocean. On the west they are traversed by the Rocky Mountains. The main characteristics of the climate and country have been already described. Towards the extreme north the cold is more intense, and Great Bear Lake is said to be frozen over for eleven months in the year. Near the Arctic Ocean trees cease to grow, and the vegetation consists only of mosses and lichen. In winter it seems to be perpetual night, and in the shortest day there are only a few minutes between sunrise and sunset. In the summer the sun seems hardly to set at all, and this is the time when the explorer and traveller has to push his journey forward.

(2) **Provisional Districts.**—*Assiniboia* contains about 95,000 square miles, and is bounded on the south by the International Boundary, on the east by Manitoba, on the north by a line drawn near 52° lat., and on the west by a line near 110° west long.

The Qu'Appelle Valley is in the district of Assiniboia, and is one of the favoured parts of the North-West. Here was the site of the famous Bell Farm, now partly broken up, which covered an area of 100 square miles, being managed by a Company. The original capital was £120,000, and in 1887 no less than 10,000 acres were under cultivation. The Bell Farm was one vast wheat-field with furrows four miles long. To plough a furrow outward and another returning was a half-day's work for a man and team. The soil of this valley is very good, and a rich black mould has been found to extend many feet below the surface. The general features of this district, especially on the eastern side, resemble those of Manitoba, but the land itself lies somewhat higher, as the second plateau or steppe is reached with its

[1] Appendix IX.

average altitude of 1600 feet. It is to this Province that a number of emigrants from Scotland and the East End of London were sent in 1883, 1884, 1885. Still later in 1888 and 1889 about 80 families of crofters from the Hebrides were settled under a system of aided emigration, each family receiving an advance of £100 to £120 to start with. Progress is very marked in this district, which will, sooner or later, attain to the dignity of a Province of the Dominion. Within twelve months a settlement can be formed and schools established on the wild and unoccupied prairies. 'The Government allowance is always liberal, and the arrangements are such that directly a district becomes even sparsely settled every child can find a school within two miles[1].' To the north of Assiniboia district, and on either side of the Manitoba and North-West Railway, the country is being rapidly developed, especially in the neighbourhood of Saltcoats and Yorkton. Here is a small colony consisting principally of small farmers or farm labourers from England, and numbering 340 souls, although the first settlers did not come till September, 1887. In the Saltcoats district there are also German and Icelandic colonies.

Alberta. This district has an area of 100,000 square miles. It lies west and north-west of Assiniboia, and is bounded on the west by the Rocky Mountains. On the south both Assiniboia and Alberta are terminated by the 49th parallel of N. lat., the International Boundary Line. Alberta has, more than any other district of the North-West, attracted English capital and become the scene of English enterprise, for here the ranches are the best in Canada. The great characteristic of the climate is the

[1] Prof. Fream, 'Agricultural Canada.' Published under the direction of the Government of Canada, 1889.

Chinook wind which blows from the west. 'The district of the Chinook wind—the country of the great cattle and horse ranches—extends from the International Boundary on the south to the Red Deer River on the north, and from the Rocky Mountains on the west to about 140 miles east. The foot-hills stretching for about twenty miles east of the mountains, are generally bare of trees, but in spring they are soft and green with the verdure of innumerable grasses, which, once the favourite food of the buffalo, are now as eagerly sought after by the cattle that have taken their place. This is the celebrated grazing country where, in the latitude of Labrador, cattle and horses range winter and summer without shelter. It is more than probable that the farmers on the east side of the Rocky Mountains, especially those of Alberta, will find a steadily growing market, particularly for meat and dairy produce, in British Columbia. From Calgary, the capital of Alberta, through the Rockies, the Selkirks, and the Gold Range to the Pacific coast is 36 hours' journey by the Canadian Pacific Railway, which should act as a line of distribution all along its route [1].'

The district of *Saskatchewan* comprises about 114,000 square miles. It lies to the north of Assiniboia and Alberta. As the branches of the great Saskatchewan pass through its valleys it has many fertile tracts which in course of time must be opened up and prove very profitable. Emigrants are now able to travel from the Canadian Pacific Railway to Prince Albert by the Manitoba and North-West Railway, and a line is in course of construction from Regina in Assiniboia, constituting two profitable arteries of trade in the future.

The district of *Athabasca* has an area of 122,000 square miles. It is bounded on the south by the district

---

[1] 'Agricultural Canada,' 1889.

of Alberta, westward by the Province of British Columbia along the 120th west longitude, on the north by the 60th parallel of latitude. The eastern boundary is, roughly speaking, the Athabasca River from the northern boundary of Alberta to Athabasca Lake, and then the Slave River to the point where it is intersected by the 60th parallel of latitude.

(3) **Rivers.**—The Mackenzie is the longest river in the great North-West Territories, and following its tributaries—the Peace River and Finlay—to the Rocky Mountains its length is about 2000 miles. This is as long as the St. Lawrence from its head-waters beyond Lake Superior. This is a country indeed of 'magnificent distances.' The Athabasca is also a tributary of the Mackenzie, rising near Mount Brown in the Rockies and flowing into Lake Athabasca. It is 900 miles long.

The *Slave River* carries the waters of Lake Athabasca to the Great Slave Lake and is 200 miles long. The Pelly, Yukon, and Great Fish River also belong to the Arctic system.

The *Saskatchewan*, which flows into Lake Winnipeg and is connected with the Hudson's Bay system, has two branches, North Saskatchewan and South Saskatchewan. The north branch is 770 miles long, the south 800 miles. The sources of the Saskatchewan are near those of the Athabasca, not far from Mount Brown and Mount Hooker in the Rocky Mountains. The Bow River also, which is an affluent of the Saskatchewan, rises near a tributary of the Missouri which flows southward into the Mississippi. Within a short distance different waters rise which flow northwards through the Mackenzie into the Arctic Ocean, westwards through the Saskatchewan into Lake Winnipeg, and thence by the Nelson into Hudson's Bay, and, southward, through the Missouri and Mississippi into the Gulf of Mexico.

The *Churchill River* rises in Saskatchewan Province and flows into Hudson's Bay a little north of the Nelson. Its length is 1100 miles.

The *Coppermine River* rises in Point Lake and flows north into the Duke of York Archipelago and the Arctic Ocean.

The *Assiniboine* rises in latitude 52° N., longitude 103° W., and after flowing southerly about 120 miles winds to the East and, joining the Red River, flows into Lake Winnipeg, and so into Hudson's Bay.

The *Albany* and *Moose* Rivers flow into the south-west side of Hudson's Bay. The Albany River is fully 450 miles long.

On the east shores of Hudson's Bay there are a few rivers flowing from the direction of Quebec and Ontario provinces, such as the Rupert, East Main Rivers — of no great size and length. To the north is Great Whale River, but the interior of the country does not resemble the great North-West. There is a large tract of country called the 'Barrens' stretching across the northern parts of the continent.

(4) **Lakes.**—*Wollaston Lake.* This lake is on the watershed and sends part of its waters northerly to Lake Athabasca and to the Mackenzie River, and part to Deer Lake on the south and so through Churchill River into Hudson's Bay.

Lake Athabasca is 200 miles long.

Great Slave Lake is situated between latitude 60° 40′ and 63° N., and longitude 109° 30′ and 117° 30′ N., its greatest length from East to West being 280 miles, and its greatest breadth 50 miles. It is frozen over for six months of the year.

Great Bear Lake. Here in the north is almost perpetual winter, the line of grasses and grains has been passed at about 65° north latitude, and this lake is partly within the Arctic Circle.

There are many smaller lakes, such as *Trout Lake* on the river Deer, a tributary of the Severn.

Severn Lake, on the Severn.

Setting Lake and *Knee Lake* on the Hayes River, flowing into Hudson's Bay near the Nelson.

South Indian or *Big Lake,* and *North Indian* or *Sandy Lake* on the Churchill River basin.

Baker Lake and *Doobaunt Lake* are connected with the Chesterfield Inlet on Hudson's Bay.

Beechy Lake, Lake Pelly, and *Lake Macdougall* are in the Great Fish River Valley.

In the vicinity of Great Bear Lake and Great Slave Lake there are numberless lakes such as Lakes Tâche, Gravelin, Sequin, Rey, Fabre, Aylmer's Lake, Artillery Lake, and many others, affording means of communication through the country from Hudson's Bay to the Rocky Mountains.

In fact, the whole vast region of the North-West is a network of lakes and rivers through which hunters and *voyageurs* have, little by little, found their way by well-known routes and 'portages.'

(5) **Towns.**—In the North-West Territories there are few towns or villages except along the line of the Canadian Pacific Railway, from the Manitoba frontier to the Rocky Mountains.

Battleford (Saskatchewan District) on the Battle River, near its junction with the North Saskatchewan, was formerly the capital of the North-West Territories, but its place has been taken by *Regina* in the Assiniboia district. Battleford is 300 miles distant from the Pacific Railway. The city of *Regina* is 356 miles west of Winnipeg. In 1882 the only sign of human occupation was three large canvas tents. Regina owes its sudden rise and importance to the Canadian Pacific Railway. It is the home of the Lieutenant-Governor of the North-West

Territories, and the meeting place of his council, and the head-quarters of the Mounted Police. The city contained in 1886 300 houses and about 1000 people. It now contains 9540 inhabitants. Here Louis Riel, who headed a rebellion of Indians and French half-castes, was executed in 1885. The region round Regina and along the Qu'Appelle Valley was a favourite haunt of the buffalo in former times. The surrounding prairie is flat and for the most part treeless.

Moose Jaw, Assiniboia, is 42 miles west of Regina, and bids fair to rival the capital itself.

Medicine Hat is a station 300 miles from Regina and 2083 miles from Montreal.

Lethbridge is on the Belly River in the south-western part of Alberta district. It is a busy town of 1000 inhabitants. From this point, 100 miles away, the great range of the 'Rockies' is visible, rising up from the immense level plains as land from the ocean. Near Lethbridge the hills are crossed, dividing the watersheds of the north and south. Within a few hundred feet the waters here flow in opposite directions. Near Lethbridge, which is connected by a branch line with the Canadian Pacific Railway, extremely valuable coal mines have been discovered, of immense advantage to the Dominion. Formerly coal had to be brought to the districts of the North-West from Pennsylvania.

Banff is a celebrated sanatorium in the midst of a great National Park in the Rocky Mountains.

Calgary, in Alberta, is 2262 miles west of Montreal, and is the great western outpost of the North-West Territories. Its beginnings do not date further back than four or five years. It is the rising town of the provisional district of Alberta, the great ranching division of the North-West.

Other important towns are Saltcoats, Broadview,

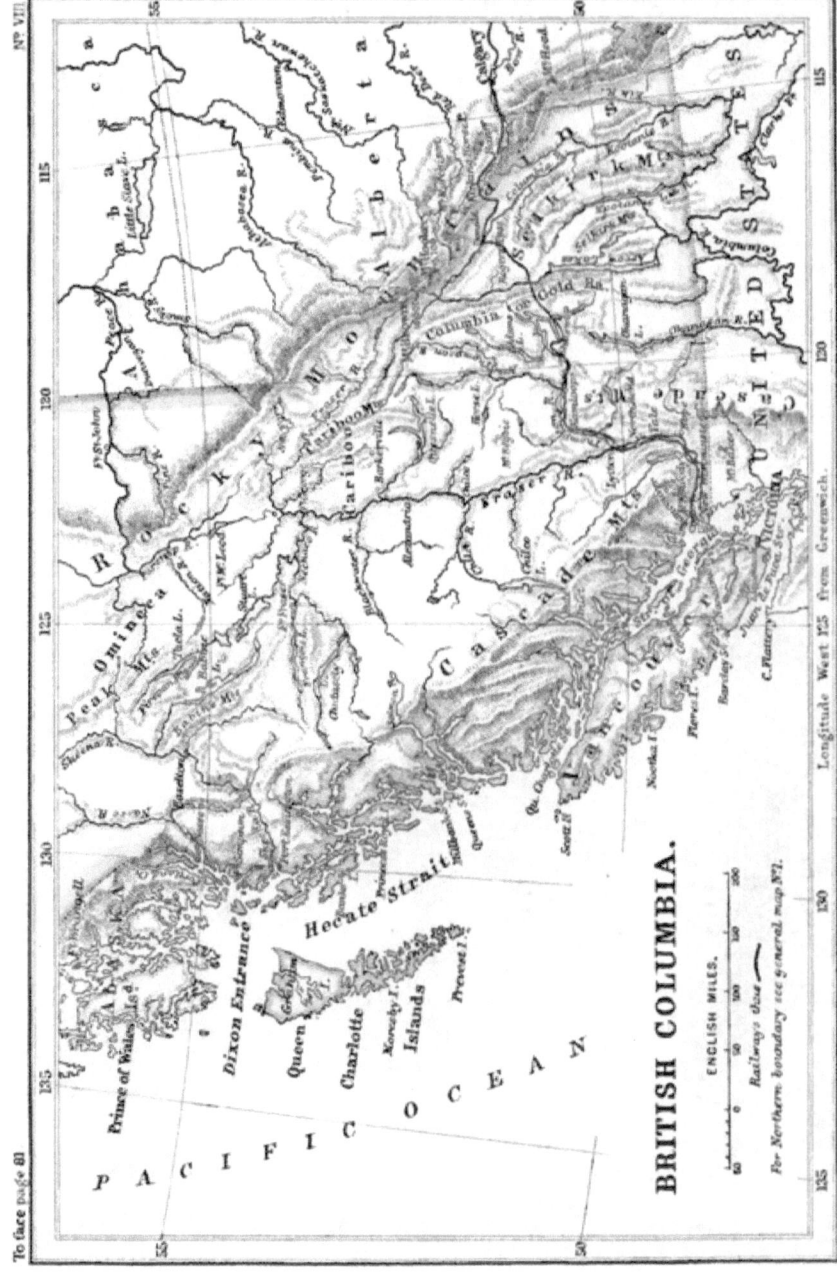

Summersberry and Indian Head, Dunmore in Assiniboia, Edmonton in Alberta north of Calgary, Prince Albert in Saskatchewan, soon to be connected by rail with the main route.

(6) In addition to the above we must notice the Forts and Factories of the Hudson's Bay Company. The most important of these are:—(1) *York Fort, Severn Factory, Haye's Factory, Rupert's Factory, Main Factory,* and *Moose Fort* on *Hudson's Bay* and *James's Bay.* (2) *Carlton House, Fort Edmonton, Fort Pitt, Cumberland House, Fort Mac Leod, Hamilton, Calgary* in the *Saskatchewan Basin.* (3) *Fort Simpson, Resolution,* and *Chippewyan* on the *Mackenzie.* The latter Fort was the starting-point of the great explorer, Mackenzie. (4) *Fort Pelly* and *Fort Ellice* not far from the trading post of Qu'Appelle, in the valley of the Assiniboine.

A tract of country in the North-West which will probably be opened up shortly is the Peace River Valley. The navigation of the river opens and closes about the same time as the *Ottawa.* In course of time the Mackenzie River Valley, with its vast forests and other wealth, may be opened up and brought into communication with the rest of the Dominion.

CHAPTER VIII.

British Columbia.

(1) UNTIL recently British Columbia, which is the most remote of the seven Confederated Provinces of the Dominion, seemed to be cut off by the huge barrier of the

Rocky Mountains, but it has now been linked to the rest by the Canadian Pacific Railway. In natural features this Province differs from the others in a remarkable degree. Once past the Rockies, a different climate is felt on the slope of land which trends to the Pacific Ocean. Warm currents of air and ocean from the west come to the coast and keep the climate, especially in the maritime districts, mild and equable[1]. The northern limits of the grains and grasses extend up to Fort Liard and to the Yukon, almost under the Arctic Circle. South of the Boundary Line a region of summer droughts is reached. The tropic currents perform for British Columbia the same service which the Gulf Stream does for the British Isles and Norway. They enable vegetation to flourish at high latitudes. Lieutenant Maury has described the physical conditions which prevail along the western shores of North America. There are, he observes, two large currents of warm water, having their beginning in the Indian Ocean. One of them is the well-known Mozambique current, called at the Cape of Good Hope the L'Agulhas current. Another of these warm currents from the Indian Ocean makes its escape through the Straits of Malacca, and, being joined by other warm streams from the Java and China Seas, flows out into the Pacific, like another Gulf Stream, between the Philippines and the shores of Asia. Thence it takes the great circle route for the Aleutian Islands, tempering climates towards the north-west coast of America. The winds also passing over its waters carry warmth in winter far inland to the Rockies. Between the physical features of this 'the Black Stream' of the Pacific, and the Gulf Stream of the Atlantic, there are several points of resemblance. Sumatra and Malacca correspond to Florida and Cuba, Borneo to the Bahamas. The coasts of China

---

[1] See p. 83 of 'Guide Book for Settlers,' Ottawa, 1882.

answer to those of the United States, the Philippines to the Bermudas, the Japan Islands to Newfoundland. As with the Gulf Stream, so also with this China current, there is a counter current of cold water between it and the shore. The climates of the Asiatic coast correspond with those of America along the Atlantic, and those of British Columbia, Washington and Vancouver resemble those of Western Europe and the British Isles.

(2) The climate of British Columbia may be classified as insular, semi-continental, and continental. The first variety is found in Vancouver, Queen Charlotte Islands, and along the broken coast-line; the second between the Cascade Range and the coast; and the last on the plateaux between the Cascades and the Rockies. The climate of the coast regions is like that of the British Isles. Here the southern wind prevails, laden with moisture from the ocean currents, resembling the Gulf Stream winds which sweep over England from the south and south-west. East of the Cascade Range, which is the lesser division of the country, the variations of heat and cold are greater, and droughts frequently prevail. But British Columbia is naturally free from the extremes of the North-West Territories, and also from the fogs and the cold mists which descend on the eastern shores of North America from Baffin Bay, and make the exposed parts of Newfoundland and Nova Scotia cold. On the Pacific side of North America there are no great reservoirs for ice and icebergs as in Baffin Bay on the east. Behring's Strait is a comparatively shallow and narrow outlet compared with Baffin Bay, and the drift of the ocean currents is *towards*, rather than *from*, the Arctic Circle. Moreover, the long projecting shores of Alaska, which are continued for a long distance by the broken Aleutian group, form a natural breakwater to shield the coasts of British Columbia from the rigours of the north.

(3) The fishing grounds of the North Pacific are especially prolific off Queen Charlotte Islands, the Skeena River and Stikeen River, and testify to the fact that the cold currents, which are best suited for fish, prevail here [1]. Following the line of the coast, the climate of Victoria in British Columbia has been described as being beyond comparison the best suited to the taste of the English on the Pacific coast. 'It has all the sun and none of the evening fogs of San Francisco; the blue sky without the rain of Portland; snow close in sight on the towering Olympia range, and yet it is never cold; hundreds of miles of inland navigation, fish at all seasons, sea and land otter, deer, elk, beaver, mink, marten, silver and sable fox, and the finest grouse shooting in the world.'

(4) **Boundaries and Area.**—British Columbia is bounded on the north by the 60th parallel of latitude, west by Alaska and the Pacific, south by the 49th parallel or United States boundary, and east by the Rocky Mountains. In virtue of a well-known decision, the line between Vancouver and Washington runs through the Haro Channel and the San Juan Archipelago. This archipelago breaks up the main channel into three straits. (1) The Rosario Strait on the east; (2) the Douglas Channel in the centre; (3) the Haro Strait on the west. The United States, therefore, gained the control of the coast and islands from the east to the Haro or Western Channel.

The area of this Province is 341,305 square miles, or nearly three times the size of the British Isles [2].

(5) **Mountains.**—The main mass of Vancouver Island is a partially submerged mountain chain, the highest peak of which, Mount Arrowsmith, rises to the height of

---

[1] Appendix X.
[2] See p. 7 of 'Official Handbook,' 1890.

nearly 6000 feet. It is continued to the south in Washington Territory and to the north in Queen Charlotte Islands. The range is precipitous and descends abruptly to the coast. On the mainland there are four ranges [1], (1) The Coast or Cascade Range, which is in reality a prolongation of the Sierra Nevada of California. It commences above New Westminster, and extends parallel with the coast as far as Mount Elias. (2) The Gold Range. (3) The Selkirks, with their extensions north and south. (4) The vast range of the Rocky Mountains.

At certain intervals there occur the following well-known gaps through the Rockies:—

Twenty miles north of the Boundary Line the Kootenay Pass traverses the Rocky Mountains. The waters of the Belly River upon the east and those of the Wigwam River upon the west have their sources in this valley 6000 feet high.

Fifty miles north of the Kootenay, the Kananaskis Pass cuts the three parallel ranges which traverse the Province running north and south. The height of land is 5700 feet.

Thirty miles to the north is the Vermillion Pass, 4900 feet.

Twenty miles further still the Kicking Horse Pass, 5210 feet [2], utilised by the Canadian Pacific Railway.

The Howse Pass, 4500 feet.

The Leather Pass (Jasper's House, or Yellow Head), 3760 feet.

The outflow of all these passes, with the exception of the Yellow Head, seeks on the east the river system of the Saskatchewan and on the west the Columbia and its tributaries. The Yellow Head on the other hand sends

---

[1] Appendix XI.
[2] See p. 342 of Hayden and Selwyn's 'North America.'

its dividing waters into the Athabasca on the east and into the Fraser River on the west[1].

(6) **Rivers.**—In the Island of Vancouver there are no navigable rivers, those that exist resembling winter torrents flowing with great velocity from the mountains and high ground of the island. On the mainland the most important rivers are :—

The *Peace River*, formed by the waters of the Findlay and Parsnip Rivers in the Rocky Mountains, near the sources of the Skeena. It finds its way through the gorges of the mountains eastward by the defile known as Peace River Pass, and flows across the great North American plains until it reaches Lake Athabasca. It was along the valley of this river that Mackenzie, the intrepid explorer of the Mackenzie River Basin, found a route to the Pacific Ocean over the Rocky Mountains. Gold is said to be found in the Peace River Valley, at Omineca and along the western slopes of the mountain range, and the whole course lies within the grain and grass-bearing zone, which reaches its highest latitude here.

Of this river Sir W. Butler has written : 'Unlike the prairies of the Saskatchewan, the plateau of Peace River is thickly interspersed with woods and thickets of pine and poplar. Its many lakes are free from alkali, and the willows affords sustenance to the moose.' The river is sharply furrowed and so deep that sometimes it seems to fill up the entire bottom of a narrow valley through which it runs. More frequently a wooded terrace lies between the foot of the ridge and the brink of the water. The soil is a dark sandy loam, the rocks are chiefly lime and sandstone, and the numerous slides and huge landslips along its lofty shores render visible strata upon strata of many coloured earths and layers

[1] Butler's 'Wild North Land.'

of rocks and shingle, lignite and banded clays in rich
succession. A black bituminous earth in many places
forces its way through rock and shingle and runs in
long dark streaks down the steep descent. Even when
April comes the river lies in motionless torpor solid
with its weight of ice four feet in thickness. Then
comes the sudden break up and the irresistible rush
of waters. Where the Peace River passes the Rockies
the valley is about two miles wide, the river 250 to
300 yards wide. The gorge is one of magnificent beauty,
with the 'glittering crowns of snow' 8000 feet above
and the 'wide beautiful valley almost filled by the
river, tranquil as a lake and bearing on its bosom, at
intervals, small inlets of green forest.' The scenery is,
perhaps, more beautiful than the dizzy glory of Shasta
and the precipices of Yosemite.

The *Skeena* flows westward into the Pacific, and is
ascended for some distance by steam vessels from Nanaimo, in Vancouver Island. It is one of the routes to
the Omineca gold mines.

The *Nasse* is at the extreme north, near the Alaska
boundary, and flows out at Observatory Inlet. It is
navigable for twenty-five miles, and its valleys are believed to abound in gold. The shortest route yet sketched
out to the Pacific is by the Yellow Head Pass, along the
valley of the Fraser River and thence northward to the
Skeena Valley, and down that valley to Port Essington,
or crossing over to the Nasse Valley and down to Fort
Simpson near its mouth on the sea. The route of steamers
would then be along Dixon Channel, north of Queen
Charlotte Island, and in the future such an alternative
route may be adopted.

The *Findlay River*, called after a fur-trader, has many
large tributaries. It is something like a huge right
hand spread out over the country, of which the middle

finger would be the main river and the thumb the Omineca. 'There is the North Fork which closely hugs the main Rocky Mountain Range. There is the Findlay itself, a magnificent river, flowing from a vast labyrinth of mountains, being unchanged in size or apparent volume 120 miles above the Forks we had left.... The Omineca lay before us stretching to the westward amid cloud-capped cliffs and snowy peaks, known to the gold seeker but not accurately. It is but one of that score of rivers which 2500 miles from these mountains seek the Arctic Sea through the gateway of the Mackenzie[1].'

The *Fraser River* rises near the Yellow Head Pass, not far from Mount Hooker and Mount Brown and the sources of the Athabasca, and flows northward first of all, until it breaks through the Cariboo mountains near Fort George. Turning suddenly to the south, it passes through that part of British Columbia which lies between the Cascades and Gold and Selkirk Ranges, and is the chief river of the district. The Cariboo plain is a promising agricultural district[2].

Great Columbia River. The Canadian Pacific Railway crosses the Rocky Mountains at 'the Kicking Horse Pass.' At the summit of this pass is a lake, whose waters drain in two directions, eastward by the Bow River into the South Saskatchewan and then into Hudson's Bay and ultimately the Atlantic, and also westward to the Columbia River, and so to the Pacific. Here, in these Alpine regions, are the sources of the Columbia River, which, after a great bend round the Selkirk Range, flows south into the Pacific in Oregon (United States) territory. It is deserving of note that the Yukon in Alaska, the Fraser, the Skeena, the Colorado, and the Columbia, are the only considerable rivers in North

---

[1] Butler's 'Wild North Land.'
[2] Appendix XII.

America west of the Rocky Mountains which find their way over the western plateaux through the gorges of the intermediate mountain ranges to the Pacific Ocean. The great loop of this river round the Selkirk Range has been called the 'Big bend of Columbia.' The region thus included is rich in minerals, timber, game, and fish. Out of its district no less than £10,000,000 worth of gold has been procured. The Canadian Pacific Railway, instead of following the course of the Columbia River, cuts across the Selkirk Range by some very skilful engineering.

The *Thompson River*, a tributary of the Fraser, upon which the town of Kamloops, meaning in Indian language 'the junction of the waters,' is founded, is a noteworthy river. In crossing the Rocky Mountains and traversing British Columbia, the railway engineers have made use of a series of deep cañons. At the summit of the Eagle Pass begins the Eagle River, and the railway follows it down till it reaches the Shuswap Lake. 'This is a most remarkable body of water. It lies amongst the mountain ridges, and consequently extends its long narrow arms along the intervening valleys like a huge octopus, in half a dozen directions. These arms are many miles long, and vary from a few hundred yards to two or three miles in breadth, and their high, bold shores, fringed by the little narrow beach of sand and pebbles, give beautiful views. The railway crosses one of these arms by a drawbridge at Sicamous Narrows, and for fifty miles it winds in and out the bending shores. This lake, with its bordering slopes, gives a fine reminder of Scottish scenery[1].' The *Eagle River* flows into the Thompson River and the Thompson into

---

[1] See p. 41 of 'A Canadian Tour,' by the London 'Times' Correspondent, 1886.

the Fraser, and it is along the valleys of these rivers that the railway is constructed.

(7) **Coast-line.**—The coast-line of British Columbia, owing to its broken character and the number of islands that adjoin the continent, is unusually long. It measures no less than 7000 miles, a distance equal to twice that of the British Isles, and is most beautiful and picturesque. This is the description given by the Earl of Dufferin in 1876 :—

'Such a spectacle as this coast-line presents is not to be paralleled by any country in the world. Day after day, for a whole week, in a vessel of 2000 tons, we threaded an interminable labyrinth of watery reaches, that wound in and out of a network of islands, promontories, and peninsulas for thousands of miles, unruffled by the slightest swell from the adjoining ocean, and presenting at every turn an ever-shifting combination of rock, verdure, forest, glacier, and snow-capped mountains of unrivalled grandeur and beauty.'

(8) **Lakes.**—In the south Lakes Kootenay, Okanagan, Upper and Lower Arrow, Shuswap, Harrison, and Lilloet. In the centre Quesnelle, Horse Fly, Cariboo, Cross, Babine, Tacla; in the north Stuart, Thutage, Dease. But of course the lake and river system of this country cannot compare with (1) the St. Lawrence; or (2) the Manitoba; or (3) the Mackenzie. However, from its natural harbour advantages and diversified coast-line there is an easy access from one end to the other, which compensates for the difficulties of travel in the interior.

(9) **Forests and timber.**—The great currents of westerly winds blowing across the Pacific, strike upon the coasts of British Columbia and Alaska. These winds are laden with moisture gathered in their course over the sea, and discharge it in copious showers upon the land. Wherever they touch North America we find magnificent

woods and forests.   More than half the area of British
Columbia is covered with the finest trees in the world.
For hundreds of miles the valleys are densely wooded, and
gigantic pines and huge cedars clothe the sides of the
mountains, even up to the regions of eternal snows.   The
king of the British Columbian pines is the Douglas fir,
which rivals the Weymouth species of New Brunswick, on
the east coast of North America.   Under the name of the
Oregon pine it is a most valuable article of commerce.
It frequently grows 300 feet high with a diameter of eight
to nine feet.   The white-cedar is a well-known forest
giant of the Fraser Valley, and on Vancouver a species
of oak grows plentifully.   Amongst other trees are the
hemlock, maple, and arbutus.   Burrard Inlet is the centre
of the British Columbian timber trade, and is only a few
miles distant from New Westminster.   It is nine miles
long, and is deep and safe.   The total annual produce is
estimated at 200,000,000 feet of timber.

(10) **Population.**—The population of British Columbia,
including Vancouver Island, is calculated to be 100,000,
including 25,000 Indians and 7000 Chinese.   However,
the country is filling fast, and in 1888 no fewer than
11,000 fresh immigrants settled there.   In the cities of
the coast large numbers of Americans are to be met
with, and the dollar is the usual currency.   The Chinese
immigrants are more numerous here than elsewhere in
Canada, and the character of a Pacific seaport is semi-
oriental.   In the interior chiefly are found the 25,000
Indians, who are kept in 'reserves' or 'locations.'

European settlers must recollect that the mainland of
British Columbia, apart from the sea-board, is divided
into three sections, each differing from the other two in
its climatic and geological features.   (1) From the Fraser
River to Yale Rapids is the 'New Westminster,' or
settled district ; (2) from Yale to Alexandria, the Simil-

kameen district; (3) from Alexandria to the Rocky Mountains, called the Lilloet-Clinton district.

(11) **Towns**.—The principal cities on the mainland are Vancouver (12,000), the terminus of the Canadian Pacific Railway, and New Westminster (5000). *Kamloops* has a population of about 700, and is the entrepot of a fine ranching district.

Yale is a settlement of about 1000, and is built on a comparatively level place amongst the mountains. Just here the railway, which has passed over 600 miles of mountain work since it entered the 'Rockies,' comes into the more fruitful and level valley of the Fraser River.

Lytton is a small town built by gold-miners at the junction of the Fraser and Thompson Rivers.

Amongst other towns and settlements on the mainland are Pemberton, William Creek, a gold centre, Hope, Quesnelly, Barkerville, Okanagan.

Before this country, which has been called, not inaptly, the 'England of the Pacific,' a great future is opening out. With its wealth derived from rich gold mines, forests, and fisheries, it must become rich and attract from all quarters an enterprising population.

Vancouver Island.

(12) **Divisions**.—1. Victoria; 2. Saanich Peninsula; 3. Tooke; 4. Cowicham; 5. Salt Spring Island; 6. Nanaimo; 7. Comox and Nelson.

The area of the island is about 16,000 square miles, or 10,000,000 acres. Of these divisions Victoria and Cowicham and the Saanich Peninsula are best suited for farming purposes.

(13) **Towns**.— *Victoria*, situated at the extreme south-east of Vancouver Island, originally the depot of the great Hudson's Bay Company, has a population of 15,000. It is

NEWFOUNDLAND.

distant 70 miles from New Westminster, and between 700 and 750 miles from San Francisco. 'Its narrow harbour, which is scarcely so large as Huskisson Dock, Liverpool, is rock-bound, and surrounded by the most charming miniature bays, exhibiting grassy knolls, and here and there evergreens in all the luxuriance of tropical foliage. A river opening out above the town invites the visitor to a boating excursion ; the fresh green of the grassy reaches which stretch into the bay, the rocky promontories, the snow-covered mountains, combine to form a landscape which deeply impresses itself upon the stranger fresh from the waves of the ocean or the sombre fir-hills of Oregon and Washington.'

Esquimault is noted for its magnificent harbour, and is distant three miles from Victoria. This harbour is thirty-six feet deep, and almost land-locked, and is an imperial naval station of the North Pacific squadron with a magnificent graving-dock for refitting.

Nanaimo (5000), sixty-five miles from Victoria, has also a good harbour. It is the headquarters of the deep-sea and whale fisheries, and is a thriving coal-mining town.

CHAPTER IX.

Newfoundland.

(1) NEWFOUNDLAND is an island off the North American coast, lying between the parallels of 46° 36′ 50″ and

51° 39′ north latitude, and the meridians of 52° 37′ and 59° 24′ 50″ west longitude.

Boundaries.—It is bounded on the north-east and south by the Atlantic Ocean, on the west by the Gulf of St. Lawrence and the Straits of Belle Isle.

Extent.—Its length from Cape Ray to Cape Norman is 317 miles, and its breadth from Cape Spear to Cape Anguille is 316 miles. With it must be reckoned the dependency of Labrador, extending from Anse Sablon in the Straits of Belle Isle to Cape Chudleigh, latitude 60° 37′.

Population.—In 1884 the population was found to number 197,539, of whom nearly 100,000 lived in the Peninsula of Avalon.

(2) The following is a description of the island given by a recent writer in the 'Nineteenth Century' Magazine, December, 1888 :—

'The island is larger than Ireland ; the greater portion of it is covered with thick and almost impenetrable forests of spruce and pine-trees, interspersed with birch, larch, and poplar. The forests give way occasionally to open spaces known locally as Barrens. They are covered with a dense carpet of mosses, which, in places, attain a depth of from one to two feet. There is a great variety of mosses, and some of them are of much beauty. Long trails of stag's-horn moss strike the eye amongst the velvety greens and deep olives, and the delicate grey and intricate tracery of the reindeer-lichen gives a pleasing contrast of colour and form. Besides mosses the barrens are rich in bilberries or hurts, partridge-berries, swamp-berries, and berries of various other kinds in extraordinary abundance . . . Innumerable lakes, or as they are called in Newfoundland ponds, are thickly dotted over the country, and though there is nothing that can be called a mountain in the island (the highest elevation being only 2400 feet), there are hills from one of which no less than 180 lakes or ponds have been counted. Large rivers traverse the island in various directions, but none are navigable for any distance, for craft larger than a canoe, as they are broken by falls and rapids and soon become shallow. The two principal rivers are the Humber, running west into the Bay of Islands, and the Exploits, which falls into Notre Dame Bay, to the north-east.'

(3) **General features.**—The geography of Newfoundland is almost entirely the geography of a sinuous and broken coast-line. Everywhere this coast-line is deeply indented with bays and inlets of the sea, and there is scarcely a village in the island which has not a close access to the sea. Perhaps no country in the world has such a coast-line in proportion to its size. In the interior there are several hills stretching across the island in a north-easterly and south-westerly direction, at whose bases are low and undulating valleys full of lakes and ponds. Here and there are isolated peaks or 'tolts' rising up to a height of 2000 feet. The highest land is on the south and west. The poorest land is on the east and south, and the best nearer the north, where the largest lakes and finest timber are found.

(4) **Climate.**—The climate of Newfoundland varies very much. At St. John's, on the extreme east, where the Peninsula of Avalon stretches far out into the Atlantic, the weather is cold and chilly, being affected by the Arctic current flowing down laden with ice from Baffin's Bay. As this icy current meets the warm waters of the Gulf Stream, it meets and creates vapour which lies upon the eastern sea-board in a dense and thick fog, making the navigation off the Newfoundland banks difficult and dangerous. The fogs rarely extend inland for a distance of more than fifty miles, and the climate on the west

coast is fairly equable, far more so than in most portions of the Canadian Dominion. The thermometer in winter seldom falls below zero, and rarely rises to 80° in summer. The mean temperature for the whole year is about 43°, the mean height of the barometer about 29° 40′[1].

(5) **Soil.**—The best soil is found at the heads of the bays and inlets, as Gander Bay, the Bay of Exploits, on the shores of Red Indian Lake, the Bays of St. George and Port-à-Port, and in the valley of the Humber River, where the best timber grows. The Laurentian system is the lowest and oldest, and spreads over two-thirds of the island; the Carboniferous is the highest series, and is found on the western side, and here and there are seams of coal. The primordial Silurian rocks are found only in isolated patches; the Huronian system prevails on the eastern part of the island. The following is the order of the formation:—(1) Carboniferous, (2) Devonian, (3) Middle Silurian, (4) Lower Silurian, (5) Huronian, (6) Laurentian. Although it does not exist in large quantities, still there is gold, found in quartz veins, and associated with iron ore; silver, both native and associated with galena, copper ores, copper in various forms, galena, plumbago, iron ores of great variety, marble and limestone, slates, and building stones of every material. Copper is perhaps the most plentiful metal.

(6) **Capes.**—On the east Cape Primavista, said to have been first sighted in 1497 by Cabot; Cape Francis, once called Cape de Portogesi, showing by its name, as well as that of Portugal Cove, that the Portuguese had fisheries here. Cape Spear, in the Peninsula of Avalon at the entrance of the magnificent harbour of St. John, on the east; Cape Race, so well-known as one of the first points of land seen by outward-bound steamers, Cape Pine on the

---

[1] See paper by the Honourable Mr. Justice (now Sir Robert) Pinsent, read before the Royal Colonial Institute, April, 1885.

south; Capes Ray and Anguille at the extreme southwest; Cape Norman on the north at the entrance of Belle Isle Strait, and on the eastern side of Pistolet Bay, Cape Bauld. In such an island as Newfoundland, with its unparalleled coast-line, the minor heads and promontories are almost innumerable.

Peninsulas.—(1) That of Avalon, with a very irregular coast-line, joined by a very narrow isthmus of a few miles in width, with Placentia Bay on one side and Trinity on the other. (2) The Northern Peninsula or Petit Nord of the French. (3) Port-à-Port on the west.

Bays.—On the eastern side, Conception, Trinity, Heart's Content, Bonavista, Goose, Freshwater, Hall, Green, Canada and Griguet Bays.

On the north, Sacred, Ha Ha, and Pistolet Bays.

On the south, Trepassey, St. Mary's, Placentia, Fortune, Connaigre, Hermitage, Despair, and White Bear Bays.

On the west, St. George's, Port-à-Port, Bay of Islands, Bonne, St. John's, St. Margaret, St. Barbe Bays.

(7) **Islands.**—On all sides of Newfoundland there are a large number of small islands, especially in Bonavista and Notre Dame Bays on the east, St. John's and the Bay of Islands on the west, and Fortune and Placentia Bays on the south. Belle Isle is the most northerly; and St. Pierre, Great and Little Miquelon, which latter belong to the French, and are used as depots for their fishing fleet, are the most southerly.

The Cabot islands are a barren group on the west side, to which a name has only been recently given by the Newfoundland Legislature, on the occasion of the erection of a lighthouse. This is the only place in Newfoundland where the name of the great explorer has been preserved.

St. John's Harbour, lying to the north of Cape Spear,

is the nearest port of call from Europe, and must be distinguished from St. John at the mouth of the St. John River, New Brunswick, with its population of 44,000. This patron saint was popular in Canada, as Prince Edward Island was originally called St. John, and north of the city of Quebec there is a St. John District.

(8) **Straits and Channels** are very numerous in Newfoundland, and they are called by many different names, e. g. tickle, gut, sound, run, passage, reach, partly according to their size, the 'tickle' or 'run' signifying the narrow, tortuous channels through which the tides run swiftly.

On the east coast the Baccalieu Tickle is between Baccalieu Island and the mainland; Chandler's Reach is at the entrance to Clode Sound, Willis Reach between Willis and Cottel's Islands, Trinity Gut between Lewis Island and the mainland in Bonavista Bay, Dildo Run between New World Island and the mainland, Long Island Tickle between Long Island, Triton and Pilley's Island in Notre Dame Bay.

On the south, Colinet Passage between Great Colinet Island and the mainland, Eastern Passage and Western Passage between Dead Islands and the mainland.

On the west, Main Gut, St. George's Bay, and Belle Isle Straits between the Great Northern Peninsula and Labrador.

Isthmuses.—(1) Avalon, between Trinity and Placentia Bay. (2) New Bay, between Bay of Exploits and New Bay. (3) The Gravels or Isthmus of Port-à-Port, connecting the Peninsula of Port-à-Port with the mainland.

Rivers.—The rivers of Newfoundland are very numerous, but none of very great importance. The Gander and Exploits rivers empty into Notre Dame Bay; St.

John's into the harbour of that name; the Dildo and Black Brook into Trinity Bay; the Humber River into the Bay of Islands. It has been said that to enumerate all the rivers and lakes in the island would fill a book, the number being prodigious. One quarter of the surface of Newfoundland is covered by fresh-water lakes and ponds. Much of the interior of the island is still unknown. It is a paradise for the fisherman and sportsman.

(9) As a field for emigration Newfoundland offers room for farm-labourers. Her soil is fertile enough to keep an agricultural population. As a mineral-bearing country she is rich and capable of development. The deposit of gypsum is enormous, and building slate, granite, limestone and marble abound. At the heads of the great lakes in the interior there is said to be 3,000,000 acres adapted for settlement and cultivation, where there are large areas of fine timber land. There are many ways in which the fishing industries can be made more productive even than they now are, and employment found for more fishermen. For fishermen, miners and agriculturists Newfoundland is a country which, only 1700 miles distant from the nearest shores of the British Islands, holds forth many attractions. Hitherto the great stream of emigrants bound for more distant Transatlantic abodes have passed it by.

(10) Newfoundland is useful to us as a nursery for seamen. Her rough weather and difficult coast navigation call into play the best kind of seamanship. The number of able-bodied fishermen is about 30,000[1]. The number of sailing vessels cleared at their ports in 1882 was 1107, their crews numbering 7200; the number of steamers cleared during the same year was 176, with crews numbering 6568. Amongst such a population as this would

---

[1] See Paper by Sir Robert Pinsent, 1885.

be found a nucleus of a strong naval force ready, in case of emergency, to defend their own or British shores.

The average annual value of the fisheries is as follows [1]:—

Cod fishery	$6,034,242
Seal fishery	1,026,896
Herring fishery	581,543
Salmon fishery	114,505
Lobster fishery	104,184
All other fish	40,000
	$7,901,370
	(or £1,646,118).

(11) **Government.**—Responsible, (1) with a Governor at the head appointed by the Crown. (2) Executive Council of seven members. (3) Legislative Council or Upper House of fifteen members, chosen for life. (4) House of Assembly of thirty-one members, elected every four years by the people. The judges are generally appointed for life.

Electoral Districts are ten in number.

In the centre, (1) District of St. John's, (2) Ferryland.

In the north, (3) District of Conception Bay, (4) Trinity Bay, (5) Bonavista Bay, (6) Twillingate and Fogo.

In the south, (7) District of Placentia, (8) of Burin, (9) Fortune Bay, (10) Burgeo and Lapoile.

These districts are not mapped out with any regularity, and little account is taken of the interior, which is virtually unoccupied. The Newfoundlanders are deprived of the control of over one-third of their coastline on the west and north by the claims of the French. The 'French shore' is the great and open grievance of the colony, and interferes greatly with the progress of the island [2].

(12) **Religion.**—There are of the (1) Church of Eng-

---

[1] Paper by Sir Robert Pinsent, 1885.
[2] Appendix XIII.

land, 69,000; (2) Wesleyan Methodists, 35,000; (3) Church of Rome, 64,000. The number of churches is about 250 [1].

Education is purely denominational, and is represented (1) by five Higher Schools or Academies, (2) Commercial and Elementary Board Schools, (3) Colonial and Continental Church Society Schools, (4) Institutions of the Christian Brothers.

The **Public Debt** of Newfoundland is very small, and does not exceed £2 per head of the population [2]. Direct taxation is unknown, and even the municipal works of St. John's and the large towns are under the Board of Public Works.

(13) **Telegraph**.—Newfoundland is noted as being the first place where the project of the submarine cable was conceived and carried out. The first idea was not a Transatlantic cable, but a short line connecting Newfoundland with Cape Breton, and so reducing by two or three days the period of communication between England and New York. The first section of the designed cable lay across the island between St. John's Harbour and Cape Race before it entered the sea. Just at this time (1852) a United States naval expedition was surveying, by means of deep-sea soundings, the bed of the Atlantic Ocean, and between Newfoundland and Ireland they discovered the existence of a plateau which seemed expressly designed for the wires of a submarine cable. After four repeated trials, the oceanic cable was successfully laid down by the Great Eastern, in 1866. The shore end of it is situated in Heart's Content Bay, a small inlet in the wider reaches of Trinity Bay, on the eastern coast. This was the beginning of the system of ocean telegraphy which, spreading in every direction over the world, and

---

[1] See Paper by Sir Robert Pinsent, 1885.
[2] Ibid.

bringing every nation within speaking distance, is effecting a revolution, not only in the general ideas of the day, but also in the operations of war, commerce and diplomacy.

CHAPTER X.

Industries, Wealth, and Social Progress.

(1) THE above are in brief outline the bare geographical details of the Canadian Dominion, and of Newfoundland, but they fail naturally to convey in themselves an adequate idea of the wealth and enterprise of the 5,000,000 people who have made these countries their home. A brief glance at the results already gained in the various departments of industries will show that the Canadians have already developed in a remarkable way the resources of their country. In other colonies, such as Natal and the Cape Colony, the presence of an overwhelming native population brings with it a large number of responsibilities and many administrative problems. In Canada there is now no native question to embarrass the politician, nor has the country to grapple, as the Republic on the south, with the negro difficulties. For in the more Southern States of the Union emancipation has brought in its wake a kind of social revolution which threatens in course of time to be of a serious nature and introduce class animosities in a free Republic. Happily for themselves the Canadians can boast of a comparatively unembarrassed poli-

tical existence. The problems they have to face are not those of a struggling and anxious community, but relate chiefly to the development of the land and its wealth.

(2) **Agriculture** is a leading industry of Canada, and to prove how much the colonists grow in excess of their own requirements we learn that in 1887 they exported £8,600,000 worth of farm produce. During this year the number of cattle alone exported was 116,274, of which 63,622, valued at £1,000,000, were shipped to Great Britain. There is no cattle disease in Canada, largely through the precautions of the Department of Agriculture. The cattle themselves are of good pedigree stock from the old country and show no signs of deterioration. The herds of Shorthorns, Jerseys, and Herefords will bear comparison with those of England, which of all countries on the globe is rightly considered to be the best adapted for cattle rearing. In dairy-farming[1], a very important branch of industry, great progress has been made of late years in Canada, and the best American cheeses are admittedly Canadian. Poultry raising has gradually been developed by a thrifty farming population, more numerous and more widely dispersed than in any other colony, as has already been pointed out, and it is calculated that poultry to the value of £21,780 and thirteen millions of eggs were exported in 1887 to the United States. The fruit and vegetables of the Dominion are too well known to need noticing here, and whilst the climate is singularly adapted to English fruits, such as peaches, apples, strawberries and currants, it is also favourable to many crops which are raised with difficulty in England. Melons ripen as a field crop, and tomatoes are so plentiful that they sell for less than 2s. a bushel, while in various localities, such as County Essex

[1] Appendix XIV.

on the shores of Lake Erie, there are vineyards resembling those of France. The account given of the common cereals shows that the Dominion is *par excellence* the country for them up to high latitudes.

(3) Generally speaking, mixed farming is carried on, the growing of grain and fruit, stock raising and dairy farming being all combined in a greater or less degree. The best immigrants undoubtedly for Canada are small farmers, with a small amount of capital ranging from £500 to £1000. There exists also in the Dominion a special Department of Agriculture, with a member of the existing Cabinet at its head; and, in 1887, legislative authority was obtained for the establishment of five government experimental farms in various parts of the Dominion to test the capacity of soil and climate—one at Ottawa, for the convenience of the Provinces of Ontario and Quebec, a central institution, one at Nappan (N. S.) for the maritime Provinces, one at Brandon for the Prairie Provinces, one at Indian Head for the North-West Territories, one at Agassiz in British Columbia, from which the most profitable results may be obtained. The Agricultural College at Guelph in Ontario has long been known as the Canadian Cirencester, and there have not been wanting men like Principal Sir W. Dawson and Professor Macoun, and others, who have by their worldwide learning contributed largely to the science of agriculture. The meeting of the British Association at Montreal in 1887, in itself a kind of Pan-Anglican conclave for the advancement of science, afforded abundant proof that the range and application of modern science, especially with reference to agricultural chemistry, was being indefinitely advanced in the Dominion. Sir Charles Lyell has written in his 'Travels in North America,' p. 5:—

'In the course of this short tour I became convinced that we must turn to the New World if we wish to see in perfection the oldest monuments of the earth's history, so far at least as relates to its earliest inhabitants. In no other country are these ancient strata developed on a grander scale or more plentifully charged with fossils, and, as they are nearly horizontal, the order of their relative position is always clear and unequivocal.' Moreover, as if to show that the wonderful prairies of the North are practically inexhaustible for the uses of man, it has been pointed out by Professor Macoun that even those plains where alkali is found, and for a long time regarded as profitless, will become the most valuable of the wheat lands as settlement progresses, the alkali being converted into a valuable fertiliser by the admixture of barn-yard manure [1].

(4) **Forestry.**—In connection with agriculture and husbandry comes the science of Forestry. To conserve the existing forest wealth of the Dominion, to thin out, plant, and experiment should be one of the chief tasks of the Department of Agriculture. Nowhere in the world has Nature been more prodigal in her gifts of the forest than in the Dominion. Although there are many durable woods in South Africa and Australia, still they require protection. In the settled parts of Australia, and in Cape Colony tree-planting has gradually become a state necessity. In the latter country, European occupation has too often been followed by veldt fires, which have destroyed in a short time and with terrific destructiveness the scanty growth of hill and valley. Consequently the surface of the earth has been exposed to the disintegrating action of sun and rain, the hills have been quickly denuded, and, literally speaking, the ribs of the earth seem to be exposed. Evaporation takes place quickly and the rain clouds become scarcer and scarcer,

[1] See 'Official Handbook,' published by the Department of Agriculture, Canada, October, 1888.

and the increased radiation from the earth is destructive to natural organisms. The water-courses run dry and droughts set in with all their concomitant miseries. In Canada the earth is sheltered and screened from the heat of the sun by means of its forests, rivers never run dry there, and the accumulated mould of ages lies many fathoms deep upon the surface. Rarely do poisonous and deadly malarious exhalations arise from the earth which are so fatal in more southern latitudes of North America and other Colonies, and the Canadian Dominion seems like a providential and beneficent extension of the temperate zone for the welfare of the northern races of Europe.

At the same time the Canadian forests are a very rich mine of wealth, always encouraging a wholesome and vigorous occupation for lumberers and axemen. It is estimated that Canada exports 700,000,000 feet of timber for the United States market—a significant fact in itself—and a similar amount for England, South America, and Australia. To carry on this trade an army of 30,000 men are employed at the rate of 5s. a day. In the Quebec Province there are 100,000 square miles of timber territory awaiting purchase. To the North-West, and especially beyond the Forks of the Saskatchewan, immense areas of forests are still to be opened up as the tide of immigration rolls forward, and the means of cheap transit are provided. The Government holds a proprietary right over these timber lands, and leases them for a general term of twenty-one years to the highest bidder. The conservation of forest wealth should be one of the first cares of Canadian statesmen.

(5) **Fisheries.**—Passing on to the fisheries we shall find that here too the Canadians have at hand an important supply of wealth which they are developing with increasing perseverance. From the Arctic regions

the currents are always bringing down a food-supply for the fish, especially along the eastern coasts. This is described as 'living slime formed of myriads of minute creatures which swarm in the Arctic seas, and are deposited in vast and ever-renewed quantities upon the fishing grounds.' In addition, the fresh-water fisheries in a country so completely honeycombed with lakes as the Canadian Dominion, are of great importance. It is approximately estimated that the value of the home consumption in 1887 was worth £2,600,000, which to £3,400,000 worth exported and sold on the Dominion markets, gives a total of £6,000,000.

(6) Further, it is calculated that the Canadians possess a fleet of 7294 vessels, 1198 steamers with a total net tonnage of 1,217,766. Assuming the average value to be £6 per ton, the value of the registered tonnage would be £7,300,000. It must be noted that about seventy per cent. of the sea-borne trade was under the British flag. On the Pacific sea-board there are not wanting signs of a growing traffic, and of trade communication with the countries of the east and with the Australasian colonies. This trade carries with it great opportunities of expansion.

As in agriculture proper, and in forestry and lumbering, this shipping industry re-acts most favourably upon the physique of the nation. It is different indeed from the existence of a South African and Australian miner, and in some respects is hard and perilous; but there is a healthfulness in the sea breezes, rough as they are at times, which toiler and digger in search of gold along a malarious belt of country can never realize. The moral effect also of Canadian industries, compelling as they do the colonists to live apart, and in widely separated regions, is healthy and sound.

(7) **Manufacture.**—Since 1878 the development of

manufactures has been more marked than during any previous period in the industrial history of Canada. The statistics of the increase in the capital invested show the advance that took place in the decade from 1871–1881 [1].

	1871.	1881.
Capital invested	£15,500,000	£33,000,000
Hands employed	187,000	254,000
Amount of yearly wages	£8,170,000	£12,000,000
Total value of articles produced	£44,100,000	£61,000,000

A partial investigation made in 1884–5 indicated that in the older provinces there had been an estimated increase of 75 per cent. in the number of hands employed (over the 1878 estimate), in the amount of wages paid, and in the capital invested, while the value of the goods produced had just doubled itself. Yet these manufactures are only in their infancy. For considering the prospect of their development it is necessary to glance at the resources Nature here again has placed at the disposal of the Canadians. In nearly every Province iron has been discovered and coal-fields of immense area are being worked in Nova Scotia—most conveniently placed on the Atlantic highway—in the North-West Territories, and also in British Columbia, here again placed most conveniently on the Pacific highway and the road to the east and the native markets. Whether for purposes of coaling war vessels or for the more peaceful object of supplying merchant steamers, the fields are equally useful. In addition petroleum is known to exist in several parts of the Dominion, and wells have been worked most profitably in Ontario where the production is very large. It is stated that in the North-West very extensive sources of petroleum have been discovered, and a railway is projected to connect them with the Canadian Pacific. What were formerly considered to be the gloomy regions

[1] See p. 19 of 'Official Handbook,' 1890.

of the north, therefore, may be gradually illuminated by light and fuel, and Nature provide for men a suitable compensation. In the North-West Territories anthracite and in all the provinces excellent peat is found. With regard to another motive power useful for the purposes of mankind, viz. that of water, it is evident that the Dominion is more favourably situated than any other colony. In the first place, perhaps no country has, from its natural configuration and extensive coast-line, greater facilities for utilising tidal action, the rise and fall of the tidal wave being, especially in the Bay of Fundy, a most remarkable phenomenon. The water-power afforded in the interior is absolutely limitless, and a glance of the mighty rush of Niagara conveys an idea of the storage of force beyond and above the Falls. The Falls of Niagara are a symbol of that Titanic and ever-present power, which can be made to subserve the services of man in the smallest as well as the greatest things. The Canadian saw-mills are at once the most extensive and best appointed in the world. It is a wonderful sight to see a log taken out of the water by an automatic process, placed in position under the saws, and speedily reduced to inch boards. This summary process of reducing to boards in a few minutes a giant pine which has taken more than a century to grow is a triumph of the mechanical skill of mankind.

(8) **Revenue and Expenditure.**—The consolidated revenue for the year ending June 30th, 1888, was made up as follows [1] :—

Customs	£4,422,000
Excise	1,216,000
Other sources	1,746,000
	£7,384,000

the expenditure for the same period being £7,343,000,

[1] See p. 17 of 'Official Handbook,' 1890.

leaving a slight balance in hand. It will be noticed that considerably more than half the whole Revenue was derived from Customs. Taxation as represented by Customs and Excise amounted in 1888 to £5,635,000, or 23s. 3d. per head, as compared with 40s. 6d. in the United Kingdom, 27s. 8d. in the Cape of Good Hope, and 61s. 8d. in Australasia. The public debt on July 1st, 1888 was £56,902,000: and deducting the assets of £9,996,000 this sum is reduced to £47,006,000, or about £9, 13s. 1cd. per head of the population. It may be remarked that at the time of the Confederation of the Provinces (1867) the net debt was only £15,145,000. Within twenty years, therefore, it has been more than trebled. But this fact need not cause any apprehension when we consider that the debt has been mainly incurred for the sake of facilitating communication between the Provinces by means of Railways, and for the general improvement of the country by reproductive public works. The country which the Canadian Pacific Railway has thrown open to colonization is practically limitless, and the new townships and settlements carry with them the germs of a greater prosperity. England has been the money market from which the colonists have borrowed, and the total amount of debt payable in England on June 30th, 1888, was £35,320,000, and the several investments for sinking funds amounted to about £4,198,000. That the Canadian Dominion investments are considered safe and secure is evidenced by the comparatively low rate of interest on which the Government can borrow. The average rate of interest paid on the gross debt in 1888 was 3·12 per cent.[1] At the time of Confederation the Provincial debts were taken over by the Dominion Government. The result of this simple transfer of liabilities to the Central Government was very satis-

[1] See p. 17 of 'Official Handbook,' 1890.

factory, this Government being able to exchange the high interest-bearing bonds of the provinces for their own bonds at a lower rate.

(9) The accumulated wealth of the country may be shown in the statistics relating to banking. The increase in business in twenty-one years, between 1868–1889, will be found to be very large [1].

	1868.	1889.
Assets	£15,500,000	£51,153,000
Liabilities	8,700,000	35,012,000
Deposits	6,500,000	24,731,000
Notes in circulation	1,600,000	6,441,000
Discounts	10,000,000	37,891,000

In 1868 there was no reserve fund, but in 1889 it reached the sum of £3,993,000. Altogether the paid-up capital invested in banking on May 30th, 1889, was £12,000,000. 'In addition to the ordinary chartered banks there are the Post Office and Government Savings Banks, the deposits in which have increased from £300,000 in 1868 to £8,000,000 in 1888, the number of depositors being now estimated at 120,000, an undeniable sign of the prosperity of the working classes in Canada since Confederation. These estimates do not include the deposits in one or two chartered Savings Banks or investments in the various Loan, Friendly and Building Societies, all of which show great developments [2].'

The condition, however, of a country which boasts of 13,000 miles of railway, an unrivalled system of canals, a mercantile marine ranking fourth in the world, and a growing population of 5,000,000 cannot but be prosperous in the highest degree.

(10) With regard to immigration, it is a well-known fact

---

[1] See p. 27 of 'Official Handbook,' 1890.
[2] An 'Official Handbook of Information,' published by the Government of Canada, 30 October, 1888.

that Manitoba and the North-West Territories are being quickly filled up both by immigrants from England and settlers from the Eastern Provinces. To get on quickly many colonists from the older and more cultivated regions of Canada follow the well-known advice, and 'go west and grow up with the country.' The Government themselves have aided and systematized immigration by every means within their power. Lands are surveyed, official descriptions given, and full information both in England and Canada supplied to the intending settler.

The township in Manitoba and the North-West Territories presupposes a complete and accurate survey of the country from the International Boundary Line northwards. The unit of the township's survey is the statute mile or section of 640 acres, all the townships being made six statute miles or sections square. The section of 640 acres is divided into half-sections of 320 acres, and quarter-sections of 160 acres. A settler may obtain a grant of 160 acres of land free on condition of three years' residence and cultivation, and payment of an office fee amounting to £2, or otherwise at the upset price of 8s. or 10s. per acre.

The townships are arranged in their running from south to north, and starting from the southern frontier which is on the International Boundary Line. These tiers are marked on the map with ordinary numerals, thus, 1, 2, 3, 4, &c. Further these townships are grouped in certain 'ranges' or large sections, running from east to west, divided by lines called 'principal meridians.' The first principal meridian starts from a point on the International Boundary Line about eleven miles West of Emerson between 97° and 98° W. longitude, and is extended northwards. The second starts from 102° W. longitude, the third from 106° W. longitude, the fourth from 110° W. longitude, the fifth from

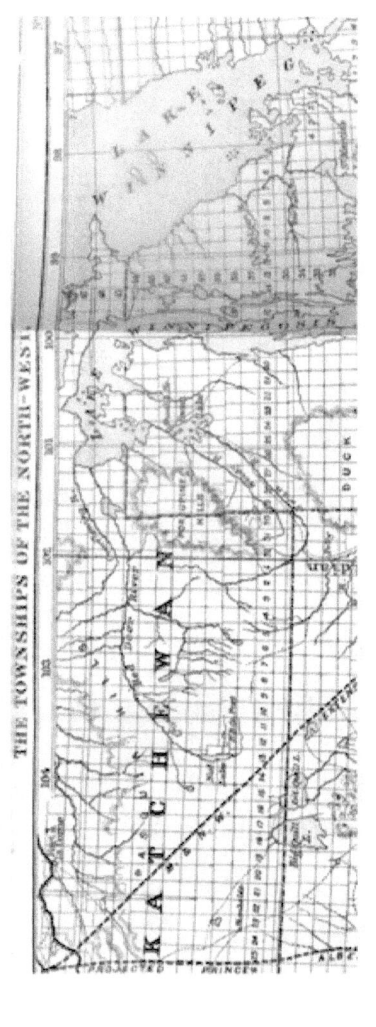

THE TOWNSHIPS OF THE NORTH-WEST

114° W. longitude, &c., &c. These ranges include therefore four degrees of longitude in each case. From the first principal meridian, which is a kind of Greenwich, and the point from which to calculate, the ranges naturally fall into two hemispheres, those on the East and those on the West. The advantages of this simple method of division, aided so materially by the level surface of the country and a wide unobstructed horizon for hundreds of miles, are numerous. In the first place there is no danger of confusion and encroachment, so often arising in new countries from the shifting and alteration of cairns and beacons. In the North-West Territories the surveyor having been beforehand with the occupier, the ordinary course of events has been reversed. The occupier can at once recognise by a reference to the map the nature and limits of his little freehold of 160 acres[1]. The surveyed lines are marked on the ground itself by iron and other kinds of monuments or posts at the corners of the divisions and sub-divisions: and as soon as the settler makes himself acquainted with these, he will instantly understand the position and extent of his own farm on the prairie, or of any other in the country. Or, when travelling in any part of the country, these posts will tell him at a glance exactly where he is, so that he cannot get lost upon any surveyed district. The Government surveyor has in fact left an accurate and trustworthy sign-post behind him, wherever the scene of his operations has been laid. Moreover, distances on the map may be calculated with very fair accuracy by the simple method of counting the number of six-mile townships East or West, North or South. The boundaries of these six-mile townships are all laid out according to the cardinal points of the

---

[1] See Map of the North-West Territories with Land Regulations, published by the Department of Agriculture, Ottawa, 1886.

compass, North, South, East, and West. The section can therefore be divided and described as having a North-East, North-West, South-East, or South-West quarter.

(11) The transfer of property, it is needless to say, is made with the utmost speed and simplicity. There are no mouldy parchments to unfold, no ancient title-deeds to consult, no greedy lawyer's fees to pay. A Deeds' Registry Office does the transfer for the colonist, and the expense of transfer to the first holder is only nominal. The emigrant does not need a very large amount of capital to begin with. The German Mennonite families began life in the North-West on an outfit of only £54 each. A sum of £150 is calculated to carry the emigrant family through the first year with a moderate degree of comfort, but they ought to be accustomed to farm work. There is no better opening in the world than here for the agricultural colonist who desires to be a freeholder as quickly as possible. With respect to the population that a township of six square miles is supposed to carry, it is clear that under the favourable conditions of soil in the North-West it will in the future be very large, even if we allow the low average of 100 persons to the square mile.

The township is a kind of parish in itself, and is the unit of colonial society. Public provision is made in these sections for the purposes of education, and naturally the section will be utilised for all purposes of local and municipal rating. It will be noticed that the principal meridian lines do not exactly tally with those of the Provinces, such as Manitoba, and those of the judicial districts, as Assiniboia; but in course of time for the sake of exactness and clearness the Provinces may possibly have their boundaries determined in the North-West by these meridian lines. Thus, the double division of boundaries, namely, that adopted to separate the Districts and Provinces and that adopted to organise and locate the constant supply of settlers, may disappear from the map altogether. The township of the Far West is in itself a wonderful example of orderly colonisation, and of nineteenth-century enterprise. It may seem uniform and stereotype, and slightly lacking in the elements of picturesqueness, but it secures comfort and plenty to the homes of thousands.

(12) Unlike the early settlements along the Alleghanies and the coasts of Maine, the victims often of religious

TOWNSHIP DIAGRAM.

```
         640 ACRES.           N.
      ┌──────┬──────┬──────┬──────┬──────┬──────┐
1 MILE│      │      │      │      │      │      │
SQUARE│..31..│..32..│..33..│..34..│..35..│..36..│
      ├──────┼──────┼──────┼──────┼──────┼──────┤
      │      │School│      │      │ H.B. │      │
      │..30..│..29..│..28..│..27..│..26..│..25..│
      │      │Lands │      │      │Lands │      │
      ├──────┼──────┼──────┼──────┼──────┼──────┤
      │..19..│..20..│..21..│..22..│..23..│..24..│
  W.  ├──────┼──────┼──────┼──────┼──────┼──────┤ E.
      │..18..│..17..│..16..│..15..│..14..│..13..│
      ├──────┼──────┼──────┼──────┼──────┼──────┤
      │      │ H.B. │      │      │School│      │
      │..7...│..8...│..9...│..10..│..11..│..12..│
      │      │Lands │      │      │Lands │      │
      ├──────┼──────┼──────┼──────┼──────┼──────┤
      │..6...│..5...│..4...│..3...│..2...│..1...│
      └──────┴──────┴──────┴──────┴──────┴──────┘
                         S.
```

persecution at home, and the victims oftener of French reprisals and cruel Indian outrages, the township of the North-West grows up in peaceful and undisturbed prosperity, the *protégé* of two paternal governments. Unlike, again, the homestead of the Virginian planter, with its dark defilement of slave labour, and the gloomy memories of the Middle Passage, the North-West township grows up with clean hands, and an untarnished heritage. No cruelty has stained its annals, no native war defaced its growth, no revolution marred its fortunes, no class hatred and religious bigotry made coarse or bitter the passions of the occupants; it rises as a full-grown community upon the level plains, full of promise and a rich prairie instinct with destiny; a second Acadia with an unencumbered heritage, and an assured position in the world.

(13) No question of naturalisation arises in connection with the emigration of British subjects to Canada, and settling in the Dominion makes no more change than a removal from York to London, Kent to Suffolk. But if an emigrant removes to the United States he has to take two oaths, one of intention and one of fact, the latter after five years' residence. The effect of these oaths is to renounce allegiance to the Queen, and surrender the British birthright. The postal system of Canada extends to every village and hamlet in the land, and is remarkably cheap. In course of time there may be a universal penny postage between the mother-country and her colonies, and a link of union created between the poorer emigrants and their relations which does not exist at present. The Money Order system is similar to that in operation in England, and the price for orders locally ranges from 2 cents (1d.) for 16s. to 50 cents (2s. 1d.) for £20. The Telegraph system is worked at a very moderate rate, some of the wires being in the hands of Government, and the telephone is used perhaps more in the Canadian than in English cities. Newspapers are to be found in almost every village of considerable size, and they are supplied with full telegraphic reports from all parts of the globe. All important news that transpires in the United Kingdom and Europe is instantly published in Canada; and, in fact, owing to the difference in mean time, an event which takes place in London at 5 p.m.

may be known in Canada at 12 noon the same day. It cannot be said, therefore, that the colonist of the North-West, if he is within range of a village centre and telegraph wire, is out of the range of civilisation.

(14) With regard to other matters the colonist will find that there is a very perfect system of Municipal Government throughout the Dominion. Both the counties and townships have their Local Councils, which regulate taxation for roads, schools, and other purposes. The burning questions of education and religion seem to have been satisfactorily adjusted in the Dominion. With regard to education every township is divided into sections sufficiently large for a school. Trustees are elected to manage the affairs, and the expenses are defrayed by local rates and grants from the Provincial Legislature. Where mixed schools, for example, of Protestant and Roman Catholic children are not possible, the law enables separate ones to be provided. Teachers are trained at normal schools at the public expense. The State assists both primary and secondary education, and there are colleges and universities open to students prepared in the lower schools. There are Schools of Medicine at Toronto and Montreal, and elsewhere, and Theological Colleges for students of Divinity. In no country in the world is general education more generally diffused than in Canada, and the highest prizes the country offers are open to all, rich and poor alike. Throughout the land the utmost religious liberty prevails, and colonists will find in most places churches and chapels already erected by men of the same persuasion with themselves. The clergy are nominated, as a rule, by the various congregations, and their stipends paid out of endowments, pew-rents, and collections. There are no tithes or church rates excepting in the Province of Quebec, where the Roman Catholic Church

holds a power over persons professing its faith. The principal religious denominations of Canada are—

Roman Catholics numbering about		1,791,982
Church of England	,,	590,537
Methodists	,,	768,608
Presbyterians	,,	697,460
Baptists	,,	291,136
Congregationalists	,,	27,000

The Roman Catholic schools owe their existence to the generosity of the Roman Catholic clergy. The professors are nearly all ecclesiastics, and are content to receive as remuneration £10 per annum. This explains the low fees paid by pupils for board and tuition, amounting to about £20 per annum. The Roman Catholic University of Laval was founded in 1854 by the Seminary of Quebec (1678), and is maintained by that body without help from the State. The two great keynotes of Canadian history are political liberty and religious toleration, and it is the carrying out these principles to their utmost under the British flag that has made the Dominion prosperous in private and public departments.

(15) Education for the professions is carried out very completely and thoroughly in the Dominion. In the Military College at Kingston, the Canadian Sandhurst, the cadets receive a careful and accurate training in accordance with the spirit of a country which in the past has proved how strong and true the martial character of its sons can be. There are still some veterans of the 1812 war alive in the country, and during the American Civil War the number of Canadian substitutes found in the Northern armies proves the aptitude of the colonists for arms, even when their services are not impressed. The supply of sailors for a colonial navy, both on the east and west coasts, is practically unlimited, and in due course of time this arm of defence will probably be strengthened and

fostered by naval depots, arsenals, and training-ships, so that by sea as well as by land the Canadian colonists will undertake the duties of self-defence, if needed, under the guidance of skilled and educated colonial officers. In a country where it is a law that military service can be demanded of all the inhabitants between the ages of eighteen and sixty, it is not likely that honourable military traditions will be allowed to die out. To prove also how valuable the fishing industry is regarded as a nursery for a navy, it is well known that the French Government have always encouraged the Newfoundland fishing fleet, under the conviction that an occupation carried on far from home, in a trying and severe climate, must give greater skill and hardihood to their sailors, and keep up their ancient spirit. From similar motives, the Admiralty at home have encouraged in past times the Polar expeditions and the search for the North-West Passage.

It is, perhaps, to the more peaceful arts of husbandry in all its branches that the Canadian Government have turned their chief attention, realising the fact that, sooner or later, the Dominion must become, by virtue of its natural advantages, one of the greatest, if not the greatest, agricultural country in the world. The Ontario Agricultural College at Guelph, established fifteen years ago, is training and sending out year by year large numbers of Canadian pupils, who by means of their careful and special education keep up the name of the Canadian farmers. In addition to this College there is the central experimental farm at Ottawa, as well as those already mentioned at Nappan in Nova Scotia; Brandon in Manitoba; Indian Head in the North-West Territory; Agassiz in British Columbia. In England we have such centres as Cirencester and Downton Colleges, where the practice and theory of farming are taught, and pupils are prepared for the careers they may have before them in the colonies. At Hollesley Bay College in Suffolk, great stress is laid upon 'Home training for Colonial life,' but it is impossible that a pupil can be equipped for his task so completely in England for Canadian farming as he would be in Canada itself. Perhaps the best plan would be, by means of some reciprocity between the Colleges at home and in the colonies, to unify the system throughout and make the curriculum and the examinations as far as possible identical, so that a pupil beginning his work in England might continue it in Canada where he left off and without loss of time. Some kind of affiliation of the Home and Colonial Agricultural Colleges for common purposes would be advantageous to all. The ground covered by the experimental farms in Canada is a wide one, as the following objects enumerated by a Canadian committee on the whole subject of agricultural teaching prove:—

(a) To conduct researches and verify experiments designed to test the relative value, for all purposes, of different breeds of stock, and their adaptability to the varying climatic or other conditions which prevail in the several provinces and in the North-West Territories.

(b) To examine into scientific and economic questions involved in the production of butter and cheese.

(c) To test the merits, hardiness, and adaptability of new or untried varieties of wheat or other cereals, and of field crops, grasses, and forage-plants, fruits, vegetables, plants, and trees, and to disseminate among persons engaged in farming, gardening, or fruit-growing, upon such conditions as are prescribed by the Minister of Agriculture, samples of such surplus products as are considered to be especially worthy of introduction. With regard to this field of enquiry, it may be noticed that most valuable experiments are now being made on the experimental farms to ascertain what kind of early-ripening wheat is best adapted to the North-West. If a sharp frost should occur on an August night, as it does occasionally in the higher latitudes, before the grain is 'out of the milk,' the sample becomes shrivelled and is deteriorated. The result of experiments is to prove that a Russian variety, called Ladoga wheat, a hard spring wheat obtained from the territory north of Riga in about 60° N., about 600 miles north of the latitude of Winnipeg, was the best, as it ripened with fewer summer days than the other varieties. Further, boreal types of cultivable plants have been extended to the Mackenzie Basin, where, besides Ladoga spring wheat, parcels of Onega spring wheat, Saxonka wheat, Petchora barley, Polar barley, Polar winter rye, and Onega oats are under cultivation [1].

(d) To analyse fertilisers, whether natural or artificial, and to conduct experiments with such fertilisers, in order to test their comparative value as applied to crops of any kind.

(e) To examine into the composition and digestibility of foods for domestic animals.

(f) To conduct experiments in the planting of trees for timber and for shelter.

(g) To examine into the diseases to which cultivated plants and trees are subject, and also into the ravages of destructive insects, and to ascertain and test the most useful preventives and remedies to be used in each case.

(h) To investigate the diseases to which domestic animals are subject.

(i) To ascertain the vitality and purity of agricultural seeds.

---

[1] 'Agricultural Canada,' p. 42. By Professor Fream. Published under the direction of the Department of Agriculture.

(*k*) To conduct any other experiments and researches bearing upon the agricultural industry of Canada which may be approved by the Minister of Agriculture[1].

(16) With regard to the question of general education in Canada, it may be gathered from what has been already said that it is cheap and popular, and admirably adapted to meet the wants of the colonist, whether living in a prairie district or in one of the crowded centres of the Dominion. The following brief sketch of the whole system will however give in outline its chief aspects:—

'In the early days of the Provinces of Ontario and Quebec, the Government set aside large areas of Crown lands, the money accruing from the sales of which have been and are applied to educational purposes. Commencing with the primary schools, which in those Provinces at any rate are free, we find them maintained partly by a school-tax levied on the lands situated in the school sections, and partly from grants from the Government which are given out of the funds above alluded to.

'These schools furnish a good English and commercial education, while recently agriculture has been added to the list of subjects taught. For this a very good text-book has been adopted. Besides these there are the high, or as we should call them, the grammar schools, in which the classics and modern languages are added to the course taught in the primary schools. Then there are the various colleges, which in a large degree were fostered by the various denominations. Each Province has several such institutions, which are officered by men of noted talent in their particular branches, and many of them by men of world-wide reputation. Some evidence of the work done by these institutions may be found in the great spirit of religious toleration which is evident in all

---

[1] See 'Canadian Lands and their Development.' Proceedings of the Royal Colonial Institute, vol. xx.

parts of Canada. The Toronto University is composed of the various sectarian colleges, including the Catholic one, and the same is the case at Winnipeg University, which is also an amalgamation of all the sectarian colleges of Manitoba, all of which work together as one University under the Presidency of the Bishop of Rupert's Land. In the Arts Faculty of Toronto no fees of any kind are exacted, and in most of the colleges of the Dominion (as in the case of the high schools where fees are exacted) they are only nominal. The Faculties in connection with Toronto University, and those of McGill University at Montreal, will compare favourably with those of the old Universities of the older world. Sir Daniel Wilson is the President of the Toronto University, Sir William Dawson of the McGill University, and Dr. G. M. Grant of the Queen's University at Kingston; and it is sufficient to mention their names to show that too much has not been claimed for the institutions over which they preside[1].'

[1] 'Canadian Lands and their Development.' Proceedings of the Royal Colonial Institute, vol. xx.

APPENDICES.

1. The three Prairie Steppes of the North-West (p. 4).
2. Louisiana and de la Salle (p. 10).
3. Herman Merivale and causes of the Dispersion and Congregation of Population in Colonies (p. 13).
4. The Prairie Soils (p. 25).
5. The Dyked Lands of Nova Scotia (p. 55).
6. The Intervale Lands of New Brunswick (p. 64).
7. The Farms of Prince Edward Island (p. 65).
8. The Rocky Mountains (p. 68).
9. Districts and Sub-districts of the North-West (p. 74).
10. Fish and Fisheries of British Columbia (p. 84).
11. Mountain Ranges of British Columbia (p. 85).
12. Agricultural Districts of British Columbia (p. 88).
13. The French Shore in Newfoundland (p. 100).
14. Dairy Farming in Canada (p. 103).

APPENDIX I.

The three Prairie Steppes of the North-West.

The following is a more particular account of the three remarkable prairie steppes reaching from Winnipeg to the Rocky Mountains, and causing the ground to rise slowly westwards :—

'The lowest and most eastern prairie level is that which includes the Red River Valley and Lake Winnipeg with its adjacent lands on the west. The average altitude of the plain is about 800 feet, the surface of Lake Superior being 627 feet above the sea; its width on the 49th parallel is only 52 miles, but its average breadth exceeds 100 miles, and its area is about 56,000 square miles of which one-fourth is water. A great part of it is more or less densely wooded, particularly that part adjacent to the lakes. The southern part, extending southward from Lake Winnipeg, includes the prairie of the Red River with an area, north of the 49th latitude, of about 6,900 square miles. This steppe is bordered on the east by the Laurentian plateau, and on the west by the first escarpment which is ascended in the neighbourhood of Macgregor, 80 miles west of Winnipeg. This escarpment, where it crosses the 49th parallel, is known as Pembina Mountain, and, trending north-west, its line is marked by the Duck, Riding, Porcupine, and Basquia Hills which lie to the west of Lake Winnipegosis. The superficial deposits of the first steppe are chiefly those of a former great lake, called Lake Agassiz, in the deeper waters of which was accumulated the fine silty material now covering the Red River Valley, and constituting a rich loamy soil of unsurpassed fertility. The valley itself is about 40 miles wide, and extends along either side of the river from north to south of the Province of Manitoba. Its surface is perfectly flat and undiversified, "the most absolutely level prairie region of America."

'When the summit of the first escarpment is reached, in the

neighbourhood of Macgregor, a vast open country called the
Great Plains, forming the second prairie steppe, is entered
upon. On the 49th latitude this second steppe is 230 miles
wide, while further north its width is not more than 200 miles.
The surface of the second prairie steppe is not so even as that
of the Red River Valley and is covered with thick deposits of
drift, consisting chiefly of detritus worn from the soft underlying rocks, but mingled with other mineral rubbish trans-
ported from a distance. From the prairie level there arise in
certain localities low hills, such as Turtle Mountain and the
Touchwood Hills, composed of accumulation of drift materials
similar to those of the Missouri Côteau on the west, the latter
being a huge glacial moraine. Turtle Mountain nowhere attains a height of more than 500 feet above the prairie. It is a
region of broken hilly ground about 20 miles square, is thickly
wooded, and hence presents a marked contrast to the general
features of the prairie. The average elevation of the second
steppe is about 1600 feet, and it is bounded westward by the
remarkable physical features known as the Grand Côteau or
hill-slope of the Missouri, which is chiefly a great mass of
glacial detritus and ice-travelled blocks, resting upon a sloping
surface of rocks of Cretaceous age, and extending diagonally
across the central region of North America, from south-east to
north-west, for a distance of about 800 miles. On the 49th
parallel, near the 104th meridian, the Côteau is 30 miles wide,
and it broadens out somewhat as it is traced northward, east of
Old Wives Lakes, to the South Saskatchewan. It is then continued to the north by a range of high lands, of which the
Eagle Hills constitute a part, to the elbow of the North
Saskatchewan . . . . The Côteau belt is particularly destitute of
drainage valleys, hence the waters of its pools and lakes are
charged with salts, particularly magnesium and sodium sul-
phates. The western part of the Côteau contains deep valleys
with tributary coulées which are mostly dry, or else occupied by
chains of small lakes which dry up in summer, and then leave
large white patches of efflorescent salts . . . . The Missouri
Côteau, which is perhaps the most remarkable monument of
the Glacial Period now existing in the western plains, is about
400 miles west of Winnipeg and fringes the eastern margin of

the third and highest prairie steppe, which extends with a gentle ascent westward, to the base of the Rocky Mountains.

'The third steppe, lying west of the Côteau, has a much thinner covering of drift deposits, a good deal of which consists of fragments of quartzite from the Rocky Mountains. Its eastern part presents in places thick deposits of true tilt or boulder clay. Its surface is more worn and diversified than is the case with the first and second steppes, and, as the Rockies are approached, it is found to consist of fragments of quartzite with softer shaly and slaty rocks and limestone. In various localities boulders are numerous, and some of these have been used in modern times by the buffalo as rubbing-stones, and are surrounded by basin-shaped depressions formed by the feet of these animals. The average altitude of the third steppe is about 3000 feet, though its eastern edge is generally a little over 2000 feet, whilst it attains an elevation of over 4000 feet at the base of the Rocky Mountains. Its area, including the high land and foot hills along the base of the mountains, is about 134,000 square miles, and of this by far the greatest part, or about 115,000 square miles, is almost devoid of forest, the wooded region being confined to a small area of its northern and north-western extension near the North Saskatchewan River and its tributaries. Its breadth, on the 49th parallel, is 465 miles, but it narrows rapidly northwards. The total area of prairie country between the parallels named, including that of all three steppes, may be estimated at 192,000 square miles. Underlying nearly the whole of the prairie region are clays, sandstones, and limestones of Cretaceous age or (in the more western parts) shales and sandstones of the Lignitic Tertiary group, the age of the latter being probably intermediate between that of the Cretaceous and of the Eocene of England. The nearest parallel to be found at home is afforded by the greater part of the counties of Norfolk and Suffolk, where Cretaceous rocks (chalk in this case) are overlaid by glacial detritus or drift.'—Extract from 'Agricultural Canada,' by Professor Fream. Published under the direction of the Department of Agriculture, Canada, 1889.

APPENDIX II.

Louisiana and de la Salle.

Louisiana was so named after Louis XIV of France, by de la Salle the French explorer, who in 1682 descended the Mississippi to its mouth. De la Salle returned to France after his discovery, and with the help and countenance of the French Court arranged the details for a French colony on the coast of the Gulf of Mexico. In July, 1684, a fleet of four vessels with 280 persons, of whom 100 were soldiers, 30 volunteers, 6 priests, set sail from Rochelle. The fate of this colony was a miserable one. The store ship was wrecked—a great loss in those days when infant settlements depended on the mother country for supplies—de la Salle failed to find the mouth of the Mississippi, and the settlement, after lingering out a miserable existence for two years, came to an end, de la Salle losing his life at the hands of his associates. In spite of its splendid beginnings, the colony was neglected by Louis XIV, who was engrossed at home in that miserable policy which led to the Revocation of the Edict of Nantes (1685). Moreover the French enterprises in the Gulf of Mexico had always incurred the enmity of the Spaniards, who regarded this sea as a *mare clausum*. It has been suggested that if any considerable number of French Huguenots had been allowed to settle in the valley of the Mississippi, with leave to worship God in the way they wished, a great French Empire might have arisen in Louisiana and Texas.

APPENDIX III.

Herman Merivale and causes of the Dispersion and Congregation of Population in Colonies.

Mr. Herman Merivale, in his lectures on Colonisation and Colonies delivered before the University of Oxford (1839, 40, 41), has pointed to several natural causes which make for concentration in new countries, p. 276. 'The causes which increase to its maximum the natural tendency to dispersion are, a wide extent of fertile soil, a wholesome climate, the absence of dense forests, and other natural obstacles, and the want of navigable rivers, upon the banks of which men are usually inclined to settle themselves in communities.' In North America it has been the presence of so much fresh water that has helped dispersion of population. It has been comparatively an easy task for the wandering colonist or pioneer to shift his home westwards, and the nature of the occupations both of fur hunter and lumberer have taken them to distant parts of the Provinces. It is a curious fact that Quebec, which commands a magnificent position on the St. Lawrence, and is the oldest city of the Dominion, should number only 65,000 out of a population of 5,000,000. Montreal (202,000) has of course largely taken her place, but the population is not centred there to the extent we might have supposed.

APPENDIX IV.

THE PRAIRIE SOILS.

'A SOMEWHAT common error made in this country, concerning Manitoba and the North-West, is to suppose that the soils covering that vast area are all alike. Such is not the case. In the Red River Valley may be seen one of the finest soils in the world, whose fertility is beyond all question. Along the route of the Canadian Pacific Railway, between Moose Jaw and the Saskatchewan River, there is a large tract of land which does not seem to offer much inducement for arable cultivation, although I was informed that it resembles the extensive sheep runs of Australia; and it is not unlikely, therefore, that it may become utilised for summer sheep-grazing, as the herbage is undoubtedly very nutritive. The area of which I speak is, in fact, the northern limit of what are called the " bad lands " of the United States; in other words, the Great American Desert. This region extends over considerable portions of the States of

Nebraska, Colorado, Wyoming and Utah, between the latitude of Santa Fé (36° N.), and that of Cheyenne (41½° N.), and between the meridians of 99° and 111° W. Much of it is occupied with the "bad lands;" and it is these, with their arid climate and scant vegetation, which impart the desert character. They also extend around the Uinta Mountains, in latitude 41°, due east of Salt Lake City, and in this locality were traversed by the original settlers in that city before their eyes were gladdened with a sight of the "promised land." It is these "bad lands"—so extensive in the States—that cross the international boundary south of the track under notice, and, rapidly decreasing in breadth, at length die out altogether as they are traced northward. But, independently of this region, there are oceans of land of excellent fertility awaiting cultivation, and I proceed to refer to the composition of some of the prairie soils of Manitoba, the remarkable richness of which arises from the accumulation for ages past of the excreta of animals, the ashes of prairie fires, and the decaying remains of plants and animals in a loamy matrix, resting upon a retentive clay subsoil. In 1884, Sir John Lawes, F.R.S., and Dr. Gilbert, F.R.S., the famous agricultural investigators of Rothamstead, Hertfordshire, published the results of some analyses they had made of four samples of Manitoba soils which they compared with typical English soils. The samples came respectively from Niverville, 44 miles west of Winnipeg; from Brandon, 133 miles west of Winnipeg; from Selkirk, 22 miles north-east of Winnipeg, and from Winnipeg itself. These soils showed a very high percentage of nitrogen; that from Niverville nearly twice as high a percentage as in the first six or nine inches of ordinary arable land, and about as high as the surface soil of pasture-land in Great Britain. That from Brandon was less rich, still the first twelve inches of depth is as rich as the first six or nine inches of good old arable lands. The soil from Selkirk showed an extremely high percentage of nitrogen in the first twelve inches, and in the second twelve inches as high a percentage as in ordinary pasture surface-soil. Lastly, both the first and second twelve inches of the Winnipeg soil were shown to be very rich in nitrogen, richer than the average of old pasture surface-soil. Commenting on their results, Sir John Lawes and Dr. Gilbert state that, whilst official records show that the rich prairie soils of the North-West are competent to yield large crops, yet, under present conditions, they do not give yields commensurate with their richness compared with the soils of Great Britain, which have been under arable cultivation for centuries. This is due partly to scarcity of labour, absence of mixed farming, burning the straw, and deficiency of manure. Three other surface-soils were examined by Sir John Lawes and Dr. Gilbert. No. 1 was from Portage la Prairie, 56 miles west of Winnipeg, and had probably been under cultivation for several years, the dry mould containing 0·2471 per cent. of nitrogen. No. 2, from the Saskatchewan district, about 140 miles from Winnipeg; its dry mould contained 0·3027 per cent. of nitrogen. No. 3, from a spot about 40 miles from Fort Ellice, might be considered a virgin soil: the dry mould contained 0·2500 per cent. of nitrogen. In general terms, these soils are about twice as rich in nitrogen as the average of the Rothamstead arable surface-soils, and, so far as can be judged, are probably about twice as rich as the average of arable soils in Great Britain.'—Extract from 'Agricultural Canada,' by Professor Fream. Published under the direction of the Department of Agriculture, Canada, 1889.

APPENDIX V.

THE DYKED LANDS OF NOVA SCOTIA.

'THE climate of Nova Scotia, and its extraordinary grass-growing capacity, at once indicate its adaptability both to grazing and dairying purposes. Its wonderful salt marshes merit a brief notice. Much of the soil along the Bay of Fundy consists of rich marine alluvium. The configuration of this bay is such that it presents, southwards to the open ocean, two coast-lines, those of Nova Scotia and the mainland, receding from each other at an acute angle; consequently, when the north-flowing tidal wave enters the bay it finds its lateral extension gradually contracted, and so its waters get piled up.

Farmers along the lower reaches of the Severn Valley in Gloucestershire (and it might be added along the banks of the Parrett in Somersetshire), will be familiar with a similar phenomenon, which there, however, occurs only with the spring-tides and produces the Bore. The tides of the Bay of Fundy spread themselves out over the adjacent coast-lands, and have there deposited marsh-soils of inexhaustible richness. In some of these saline swamps marsh grass grows abundantly and yields a heavy crop. But large areas of the salt marshes have been reclaimed by means of mud dykes, so built as to prevent the irruption of the tidal water, and it is these dyke lands which constitute so interesting and peculiar a feature in Nova Scotia along the Bay of Fundy and around the Basin of Mines, and on the adjacent shores of New Brunswick. The eastern dykes are strong and broad, six to eight feet high; and the land within them is hard and dry, and produces an abundance of coarse but nutritious herbage. Year after year will these reclaimed marsh lands give upwards of two tons of hay per acre, and show no signs of running out. The Salt Hay, as it is termed, costs about £1 per acre to make, and is worth from £5 to £6 per ton in the market. The cost of reclaiming and dyking these salt marshes varies between £1 10s. and £4 per acre .... In no country of America is fruit-growing better understood or more successfully practised than in Nova Scotia. The magnificent apple orchards of the Annapolis Valley stand, perhaps, unrivalled.'—Ibid., pp. 9, 10.

APPENDIX VI.

The Intervale Lands of New Brunswick.

'One of the most interesting natural features of the Province of New Brunswick is afforded by the intervale lands, that is, lands lying between the slopes of the valleys. Respecting these Professor Sheldon writes: "The *intervale* lands are, as the name suggests, found in the valleys. The name is peculiarly appropriate and expressive. In England we should call them bottom-lands or alluvial soils. They are, in fact, alluvial soils to all intents and purposes, with this peculiarity—they are still in process of formation. In some cases these *intervale* lands consist of islands in the rivers, and there are many such in the magnificent river St. John; but for the most part they are level banks on each side of the river, in some cases several miles wide and reaching the foot of the hills, which form the natural ramparts of the valley they inclose. These *intervale* lands are rich in quality and the grass they produce is very good."'—Ibid., p. 11.

APPENDIX VII.

THE FARMS OF PRINCE EDWARD ISLAND.

'THE most recent agricultural statistics relating to Prince Edward Island are those obtained in the Census of 1881, in which year there were 16,663 owners of land, 13,629 occupiers of land, 1,126,653 acres occupied, and 596,731 acres improved. The live stock comprised 31,335 horses, 90,722 cattle, 166,496 sheep, and 40,181 pigs. The population at the time was 108,891, and is now about 119,000. The agricultural produce in 1881 included 3,538,219 bushels of oats, 540,986 of wheat, 119,368 of barley, 90,458 of buckwheat, 3,169 of peas and beans, 15,247 of timothy and clover seed, and 143,791 tons of hay; also 1,688,690 lbs. of butter, and 196,273 lbs. of cheese. There is a considerable export trade in horses, cattle, and sheep to other parts of Canada and to the New England States. Perhaps the most peculiar feature in the farming of the island is the extent to which the mussel mud of the rivers is used as manure. The mud is obtained by a dredging machine, worked by horse-power on the ice over the beds of nearly all the rivers where oyster and mussel deposits occur. These deposits are from ten to thirty feet thick, and are made up of oysters, mussels, decayed fish, and sea-weed. Used as a fertiliser, this material acts promptly and effectively and produces very large crops of hay. Improved farms can here be bought at £4 per acre.

Professor Macoun, naturalist to the Geological and Natural History Survey of Canada, writes thus: "Prince Edward Island is a lovely spot and resembles very much an English landscape, as instead of the log fences used in Ontario there are hedges of English and Canadian hawthorn along the roadsides and round the farms. The houses, too, resemble the English farm houses. The soil is of sandy clay or sandy loam, red in colour. The farmers never have droughts on the island, but the seasons are very late. The ice in the Gulf of St. Lawrence during April and May renders the spring late, while the autumn is long in consequence of the heating of the Gulf waters during summer. Growth starts late in spring, is slow in summer, and has a long time to mature in autumn. As a result, there are no absolute failures of crops. The island has the best pasturage land on the continent of America."'—Ibid., p. 12.

APPENDIX VIII.

THE ROCKY MOUNTAINS.

'THE western boundary of the Prairie Region is formed by the magnificent natural rampart of the Rocky Mountains, which often present to the east almost perpendicular walls of rock, though the junction of plateau and mountain is usually flanked by foot hills, such as those to the south or west of Calgary, among which the cattle ranches of Alberta have been established. In this superb mountain range the loftiest peaks are clad with perpetual snow, thrown into bold relief by the dark green hues of the pine trees which clothe the lower slope. The spectacle of this steep straight line of snowy peaks is said to surpass the view of the Alps from the Milan Cathedral, or that of the Pyrenees from Toulouse. The range seems to culminate between the 51st and 52nd parallels about the head waters of the North Saskatchewan, and to the north gradually decreases in elevation till on the borders of the Arctic Ocean it is represented by low elevations. A common impression is that the whole of the mountainous region between the western boundary of the prairie and the coast of the Pacific is constituted by the Rocky Mountains. Such, however, is not the case, the Rocky Mountains being only the eastern portion of this region. Parallel with them run the Gold Range, and, further west, the Coast Range. From the western edge of the prairies to the Pacific, between the 49th and 56th parallels, the average breadth is 400 miles. The Rocky Mountain range of Canada is narrower than it is further south, in the United States the average breadth near the boundary line is about 60 miles, which further decreases near the Peace River to 40 miles. True glaciers appear only about the head waters of the Bow, North Saskatchewan, and Athabasca. On the western, or Pacific, side the Rockies are defined by a very remarkable and straight and wide valley, which can be traced uninterruptedly from the 49th parallel to the head waters of the Peace River, a distance of 700 miles.

'The Rockies therefore must be distinguished from the Gold Ranges, which include the subsidiary ranges called the Selkirk, Purcell, Columbia and Cariboo Mountains. The Selkirks also have more rounded and flowing outlines, but are very difficult to penetrate, the forests being extremely dense and tangled. It has been said that the younger Verendrye was the first European traveller to see the Rocky Mountains, but his description of the mountains he saw is hardly glowing enough, and certainly not what we should have expected, if his eye really rested on those magnificent ramparts. Others have claimed the honour for Clark and Lewis (1806).'—Ibid., p. 23.

APPENDIX IX.

Districts and Sub-districts of the North-West.

The following is a Census of the three provisional districts of the North-West Territories (August 24, 1885):—

District.	Sub-district.	Population.
Assiniboia	Broadview	8,367
	Qu'Appelle and Regina	9,540
	Moose Jaw	2,616
	Swift Current	363
	Maple Creek	465
	Medicine Hat	732
Saskatchewan	Carrot River	1,770
	Prince Albert	5,373
	Battleford	3,603
Alberta	Edmonton	5,616
	Calgary	5,467
	McLeod	4,450
		48,362

Ibid., p. 39.

APPENDIX X.

Fish and Fisheries of British Columbia.

'The fishery products of the Province are already remarkable. considering the small population yet engaged in the trade. The exports of fish and of fish products from Victoria alone, in the year ending June, 1888, were £279,000 in value, and the total yield, including the consumption by Indians, is over £1,000,000. The salmon of British Columbia are famous. There are twenty-one factories for making canned salmon, twelve of them being on the Fraser, and their annual out-put is from 150,000 to 200,000 cases (each containing forty-eight 1 lb. tins), with 4000 to 5,000 barrels of salt salmon. The take of salmon from the Fraser is over 8,000,000 lbs., exclusive of what the Indians procure. Fresh salmon, as well as tinned salmon, are now being shipped frozen to the markets of Eastern America and England. A remarkable British Columbia fish is the oolachan, or candle fish. It is smaller than a herring, and so oily that when dried it will burn like a candle. They are caught chiefly in the Nasse and Fraser Rivers. The fish begin running in the Nasse about the last day of March, and enter the stream by the million for several weeks. Another fish destined to be of great commercial value is the skil or black cod, which is caught in 150–300 fathoms of water and at some distance from the shore. It is found in countless numbers between Vancouver and Queen Charlotte Island. The sealing industry is productive, the catch of seals in 1887 being valued at £47,000. There are disputes however between the Canadian sealers and the Alaska Commercial Company.'—See 'Canada, a Memorial Volume,' Montreal, 1889.

APPENDIX XI.

Mountain Ranges of British Columbia.

'After the ramparts of the Rocky Mountains are passed there are the Gold Ranges, known more particularly as the Selkirk, Purcell, Columbia, and Cariboo mountains. The width of the Gold Range is about 80 miles, but north of the Cariboo district, above the head waters of the Peace River, it dies away. Between the Gold and Coast, or as they are sometimes called the Cascade Ranges, is the interior plateau of British Columbia, with an average width of 100 miles and an elevation of 3500 feet. Its height increases southward, but declines northward at the sources of the Peace River. It is dissected by deep and trough-like valleys. In the north it is closed at about latitude 55° 30′ by several intercalated mountain ranges.

'The Coast, or Cascade Range, have an average width of 100 miles. These mountains are extremely rugged, and receive on their seaward slopes the moisture from the sea, and have a very luxuriant vegetation. Vancouver Island, and the Queen Charlotte Islands to the north-west, are constituted by another parallel series of mountains, the Vancouver Range which is continued southward in the Olympic Mountains, and northward in the peninsular portion and islands of Alaska. The highest mountain in Vancouver Island is 7844 feet.'—'Agricultural Canada,' by Professor Fream, pp. 52-53.

APPENDIX XII.

Agricultural Districts of British Columbia.

'British Columbia cannot be called an agricultural country throughout its whole extent, and the province will probably be supplied with produce for some time from Alberta and Assiniboia. Its forests are very valuable, and in 1888 the exports of timber were valued at £115,000, as compared with £30,000 in 1881. Yet the agricultural resources are very great. In the Cariboo district there is a plain 150 miles long and 60-80 miles wide, and between the Thompson and Fraser Rivers there is an immense tract of arable and grazing land. The hills and plains are covered with bunch grass, on which the cattle and horses live all winter, and its nutritive qualities are said to exceed the celebrated blue grass and clover of Virginia. Between 5000 and 6000 square miles of the Peace River prairie land is within this province. Besides the mainland, there are on Vancouver Island about 1,000,000 acres of land well suited to agriculture, and on Queen Charlotte Islands about 100,000 acres, most of this being now covered with dense forests. Further, it should be mentioned that from 1858-1885 about ten millions of pounds sterling was yielded by the Gold mines, also that extremely rich coal deposits exist on Vancouver Island, and at Burrard Inlet and Nicola Valley in the mainland.'—Proceedings of the Royal Colonial Institute, vol. xviii. p. 195, 1886-7. Paper by the Bishop of New Westminster.

APPENDIX XIII.

THE FRENCH SHORE RIGHTS IN NEWFOUNDLAND.

BY the Treaty of Ryswick, in 1697, various places on the south coast, of which Placentia was the French capital, were kept in the possession of the French. In 1702 they held nearly the whole island in their own hands. Placentia was almost as firm a stronghold of the French as Port Royal in Nova Scotia. The Treaty of Utrecht secured the complete sovereignty of England, but it conceded certain Fisheries Rights which have been a continual source of trouble. The French sailors were allowed the right of catching fish, and drying them on land from Bonavista to the eastern shores of Newfoundland, and thence northward to Point Riche on the western shore. These rights have been left much as they were then defined. The Treaty of Versailles in 1783 and the Treaty of 1814 touched them, but did not alter them materially *except in* changing the limits, so that they now commence at Cape St. John on the east coast on latitude 50°, and extend north and west to Cape Ray, the south-west point of Newfoundland. The anomaly still remains that, although England's sovereignty is acknowledged, the French claim exclusive rights of Fishery along that shore.

APPENDIX XIV.

DAIRY FARMING IN CANADA.

'THE value of the cheese exported has more than doubled within recent years, Canadian cheese being now recognised as the best made in America; and of late years it has competed successfully with the English-made articles. The following figures tell the progress of this trade in eleven years:—

	Quantity Exported.	Value.
1874	24,050,982 lbs.	£700,000
1884	69,755,423 ,,	1,400,000
1887	73,604,448 ,,	1,400,000
1888	84,173,267 ,,	1,800,000

Appendix XIV.

'Such a rapid development in the cheese trade has naturally had the effect of limiting the production of butter: but nevertheless 4,415,381 lbs., of the value of £160,000, were exported in 1888.

'In September, 1888, there were in the Province of Ontario more than 750 cheese factories, using up the milk of 250,000 cows. At the same time there were 40 creameries at work. There are three associations, called the (1) Dairymen's Association of Western Ontario, (2) of Eastern Ontario, (3) the Creameries Association of Ontario. The average quantity of milk used per annum in the factories is 661,147,200 lbs., yielding about 28,660 tons of cheese, the value of which is £1,247,300, or 4·85 pence per lb. The farmers who send milk to the factories are called "patrons." The average annual number of patrons was 24,207, and of cows 148,560, the average per factory being 56 patrons, and 341 cows. Short-horns and their grades predominate in Western Ontario, but the Ayrshire is the favourite along the St. Lawrence.'—Extracts from Professor Fream's 'Agricultural Canada,' and Official Handbooks, 1889.

GENERAL INDEX.

A.

Acadia, 5, 9, 52, 116.
Africa, Equatorial, 16.
— South, 46.
Agassiz, 104.
Agriculture, Canadian, 103, 121.
Ainslie Lake, 54.
Alaska, 11, 83, 84, 87, 88.
Albany River, 78.
Albert County, 61.
Alberta District, 75, 76, 80.
Alberton, 66.
Aleutian Islands, 82, 83.
Alexander, Pope, 6.
Alexandria, 91.
Algonkins, 58.
Allagash, 59.
Alleghanies, the, 10, 32, 115.
Allumet Island, 30.
Altamaha River, 9.
Amadas, Captain, 8.
Amazon, 6.
America, South, 38.
Amherst, 56.
Anguille, Cape, 94, 97.
Annapolis, 8, 52, 56.
— County, 55, 57.
— River, 53. 54.
Anne, Queen, 52.
Anse Sablon, 28, 94.
Anticosti, 5, 29, 30.
Antigonish, 56.
Appelachians, 32.
Arctic Circle, the, 16, 19, 78, 82, 83.
— Current, 95.
— Ocean, 74, 77, 78, 88.
Argenteuil County, 31, 35.
Arichat, 57.

Army, Canadian, 27.
Aroostook, 59.
Arrow Lake, 90.
Arrowsmith, Mount, 84.
Artillery, Canadian, 27.
— Lake, 79.
Ashburton Treaty, 11.
Asia, 82.
— Central, 24.
Assiniboia District, 68, 74, 81.
Assiniboine River, 14, 70, 73, 78.
Athabasca District, 71, 76.
— Lake, 16, 19, 77, 78, 86.
— River, 77, 86, 88.
Atlantic, the, 11, 20, 24, 52, 53, 83, 88, 94, 95.
Aurora Borealis, 23.
Australia, 12, 38, 105.
Avalon Peninsula, 94, 95, 97, 98.
Aylmer's Lake, 79.
Azores, 6.

B.

Babine Lake, 90.
Baccalieu Tickle, 98.
Baddeck, 57.
Baffin's Bay, 24, 83, 95.
Bagot County, 35.
Baie St. Paul, 72.
Baie Verte, 54, 60, 62.
Baker Lake, 79.
Banff, 80.
Banks, Savings, 111.
Barens River, 70.
Barkerville, 92.
Barley, Canadian, 18, 19.
Barré, Charlotte, 39.
Barrens, the, 78.
— — Newfoundland, 94

Barrington, 58.
Bathurst, 61, 62, 63.
— Bay, 59.
Batiscan River, 35.
Battle River, 79.
Battleford, 79.
Bauld, Cape, 97.
Bear Lake, the Great, 16, 78, 79.
Beauce County, 35.
Beauharnois, 39.
— County, 35.
Bedeque Harbour, 65, 66.
Bedford Basin, 54.
Beechy, Lake, 79.
Behring's Strait, 83.
Belfast, 56.
Bell Farm, 74.
Belle Chasse County, 35.
— Isle Straits, 14, 28, 33, 45, 94, 97, 98.
Belly River, 80, 85.
Belœil Mountain, 30.
Bend of Columbia, 89.
Bermudas, 83.
Berthier, 40.
— County, 35.
Betsiamite River, 35.
Big Lake, 79.
Bird Rocks, 29.
Birds, migratory, 23.
Birtle, 73.
Black Brook, the, 99.
— Stream, the, 82.
Blackfriars, Ontario, 47.
Blizzards, 15.
Blomidon, 52.
— Cape, 54.
Blue Mountains, 43.
Bonaventure County, 35, 63.
Bonavista Bay, 97, 98.
Bonne Bay, 97.
Boston, 20.
Boularderie Island, 54.
Bow River, 77, 88.
Brandon, 73, 104.
Brant County, 49.
Brantford, 47.
Bras d'Or Lake, 54.
Briar Island, 54.
Bridgwater, 58.
Brion Island, 29.
Bristol Channel, 53.

British Association, the, 104.
— Isles, 16, 24, 38, 83, 84.
Broadview, 80.
Brome County, 35.
Brown Mountain, 15, 77, 88.
Bruce County, 49.
Bureau of Industries, 44.
Burgeo District, 100.
Burin District, 100.
Burrard Inlet, 91.
Butler, Sir W., 86.

C.

Cabot, 1, 6, 7.
— Islands, 97.
Calais, 37.
Calgary, 76, 80, 81.
California, 85.
Calumet Island, 30.
Camatha, 5.
Campbellton, 62.
Canada, Dominion of, 11, 16, 25, 27.
— Bay, 97.
— Proper, 5, 9, 11.
— Upper, 42.
Canadian people, 26.
Canals, 33, 48.
Canso, 52, 58.
— Gut of, 53, 54.
Cape Breton, 10, 52, 54, 57, 101.
— Colony, 102, 105, 110.
— of Good Hope, 82.
Capes of Nova Scotia, 54.
Capetown, 14.
Caraquet Bay, 60.
Cardigan Bay, 65.
Cardwell County, 49.
Cariboo Island, 54.
— Lake, 90.
— Mountains, 88.
— Plain, 88.
Carleton County, 61.
Carlton House, 81.
Cartier, Jacques, 5, 6, 9, 37.
Cascade Mountain, 83, 85, 88.
Cascapedia, 60.
Cataraqui River, 45.
Catholics, Roman, 40, 57, 61, 101, 117, 118.
Cavalry, Canadian, 27.
Census of Quebec, 40.

General Index.

Chaleurs, Bay des, 29, 32, 58, 59.
Chambly County, 35.
Champlain, Samuel, 37.
Champlain County, 35.
— Lake, 30, 35.
Chandler's Reach, 98.
Charles I, 52.
Charlevoix County, 35.
Charlotte County, 61.
Charlottetown, 12, 66.
Charters, Royal, 7, 8.
Châteauguay County, 35.
Chatham, New Brunswick, 62.
— Ontario, 47.
Chatte Cape, 14, 32.
Chaudière River, 4, 32, 35.
Chedabucto Bay, 54, 58.
Chesterfield Inlet, 79.
Chicago, 19.
Chicoutimi, 34, 39.
— County, 35.
Chignecto Bay, 52, 54, 59.
— County, 54, 58.
China Seas, 82, 83.
Chinese, the, 91.
Chinook winds, 76.
Chippewyan Fort, 19, 20, 81.
Christiania, 57.
Chudleigh Cape, 94.
Churchill River, 78, 79.
Cipango, 5.
Cirencester, 119.
Climate, Canadian, 16, 23.
— of British Columbia, 83.
— of Manitoba, 71.
Clode Sound, 98.
Coal Mines, Canadian, 52, 56, 80, 108.
Cobequid Bay, 54.
— Mountains, 53, 58.
Cocagne River, 59.
Colchester, Ontario, 47.
Colebrooke, 61.
Colinet Island, 98.
— Passage, 98.
Colorado River, 88.
Columbia, British, 11, 13, 14, 27, 68, 76, 77, 81.
— River, 88, 89.
Columbus, Christopher, 5, 6.
Comox, 92.
Company, London, 8.

Company, Plymouth, 8.
Compton County, 35.
Conception Bay, 90, 101.
Connaigre Bay, 97.
Connecticut River, 9.
Coppermine River, 78.
Cornwallis, 58.
— River, 54.
Cortereal, 7.
Cottel's Island, 98.
Couchiching Lake, 49.
Council of Plymouth, 8.
Councils, Local Canadian, 117.
Counties of Cape Breton, 57.
— of Manitoba, 72.
— of Nova Scotia, 55, 56.
— of Ontario, 49.
— of Quebec, 35, 36.
Cowicham, 92.
Crop averages, 18.
Cross Lake, 90.
Crossways Lake, 32.
Cuba, 83.
Cumberland Basin, 54, 60.
— House, 20, 81.
Currents, ocean, 23, 24.

D.

Dakota, 15, 17.
Dalhousie, 61, 62.
Darkies Lake, 54.
Dartmouth, 58.
Dauphine, Cape, 54.
Davis's Straits, 24.
Dawson, Sir W., 104, 123.
Dead Island, 98.
Dease Lake, 90.
Debt, Canadian, 110.
— of Newfoundland, 101.
Deer Lake, 78.
— River, 79.
De Monts, 8, 58.
De la Peltrie, 39.
— la Roche, 9.
— la Roque, 7.
— la Salle, 10.
Delaware River, 9.
Deserts of United States, 17, 20.
Despair Bay, 97.
Des Esquimaux River, 29, 35.
Detroit, 33.
Dickens, Charles, 46.

Digby, 58.
— Gut, 54.
Dildo Run, 98.
Dixon Channel, 87.
Doobaunt Lake, 79.
D'Or Cape, 54.
Dorchester, 61, 62.
Douglas Channel, 84.
— fir, 91.
— Lord Selkirk, 67.
Dover Port, Ontario, 47.
Downton, 119.
Duck Mountain, 68.
Dufferin, Lord, 68, 90.
Du Lièvre River, 34.
— Moine River, 34.
— Nord River, 34.
Dunmore, 81.
Durham County, Ontario, 49.

E.

Eagle Pass, 89.
— River, 89.
East Main River, 78.
— River, 53.
Edmonton, 81.
Education in Dominion, 118, 121, 122.
— in Newfoundland, 101.
Edward Lake, 32.
Eel Lake, 60.
Egmont Bay, 65.
— Cape, 54.
Electoral Districts, 36, 50, 100.
Elgin County, 49.
Elias, Mount, 85.
Ellice, 81.
Emerson, 73, 112.
Engineers, Canadian, 27.
England, 16, 24, 83.
— of the Pacific, 29.
Eozoon Canadense, 31.
Erie Lake, 11, 16, 33, 42, 47, 48.
Esquimault, 5, 93.
Essex County, 49, 72.
Essington, Port, 87.
Eternity, Cape, 34.
Europe, 11, 21.
Evangeline, 53.
Exploits Bay, 95, 96.

F.

Fabre Lake, 79.
Factories of the Hudson's Bay Company, 81.
Falls of the Chaudière, 35.
— of Montmorency, 34.
— of Niagara, 46.
— of St. John, 59.
Farnham, 39.
Fear, Cape, 6, 8.
Ferryland County, 100.
Findlay River, 77, 86, 87, 88.
Fish River, Great, 77, 79.
Fisheries, Dominion, 14, 22, 106.
— New Brunswick, 62.
— Newfoundland, 100.
Florida, 83.
— French, 5, 9.
— Spanish, 8.
Fogo, 100.
Forests, Canadian, 21, 22, 23, 105.
Fort Garry, 73.
Fortune Bay, 97, 100.
France, 31, 119.
Francis I, 6.
Francis, Cape, 96.
Franklin, 4.
Fraser River, 14, 86, 87, 88, 89, 90, 91, 92.
Frazerville, 39.
Fredericton, 59, 61, 62, 63.
French Lake, 60.
— River, 44.
— Shore, the, 100.
Freshwater Bay, 97.
Frontenac, 45.
Fundy, Bay of, 11, 52, 53, 58, 59, 60, 61, 62, 109.

G.

Gabarus Bay, 51.
Gagetown, 61.
Gander Bay, 96.
Garden of Canada, 43.
Gaspé County, 36.
— Peninsula, 6, 29, 32, 41.
Gatineau River, 34.
George Bay, 54.
Georgetown, 65, 66.
Georgian Bay, 49.
Germans, the, 58, 69, 75.

General Index.

Gilbert, Sir H., 6, 7.
Gladstone, 73.
Glasgow, 56.
Gloucester County, New Brunswick, 61.
Gold Range, 76, 85, 88.
Goose Bay, 97.
Grains and Grasses, 18.
Granby County, 54.
Grand Falls, New Brunswick, 61.
— Lake, New Brunswick, 60.
— Lake, 32
— Passage, 54.
— Pré, 52.
Grant, Dr. G. M., 123.
Gravelin Lake, 79.
Great Bear Lake, 74.
— Britain, 10.
— Eastern, the, 101.
Green Bay, 97.
— Mountain, 32.
Grey County, 49.
Griguet Bay, 97.
Guelph, 47, 104.
Gulf Stream, 24, 55, 82, 95.
Guysborough, 56.

H.

Ha Ha Bay, 97.
Haldimand County, 49.
— House, 35.
Haliburton County, 49.
Halifax, 5, 12, 16, 20, 56, 57, 62, 63.
— Harbour, 54.
— Lord, 57.
Hall Bay, 97.
— River, 29.
Hamilton, Fort, 81.
— Port, 12, 47.
Hampton, 61.
Hants County, Nova Scotia, 55.
Haro Channel, 84.
Harrison Lake, 90.
Harvey, 62.
Hatton County, 47, 49.
Havre, 37.
Hay, Canadian, 18.
Hayes Factory, 81.
— River, 79.
— route, 70.
Heart's Content, 97, 101.

Hebrides, 75.
Hennepin, Father, 46.
Henry, Fort, 45.
Herefords, 103.
Hermitage Bay, 97.
High Bluff, 72.
Highlanders, 66.
Hillsborough Bay, 65.
Hochelaga, 39.
Hollesley Bay College, 120.
Hooker, Mount, 15, 77, 88.
Hope, 92.
Hopewell, 61, 62.
Horse Fly Lake, 90.
Howse Pass, 85.
Hudson River, 8, 9.
Hudson's Bay, 4, 10, 70, 77, 78, 79, 81, 88.
— Company, 73, 92.
— Territory, 67.
Hull, 39.
Humber, the, 95, 96, 99.
Humboldt, 21.
Huntingdon County, 36.
Hurlbert, 69.
Huron County, 49.
— Lake, 11, 16, 33, 47, 48, 49.
Huskisson, 93.

I.

Iberville, 39.
— County, 36.
Icelanders, 69, 75.
Illinois, 16.
Immigrants to New Brunswick, 63.
Immigration, 112.
Indian Head, 61, 104.
— Ocean, 82.
Indians, the, 47, 61, 91.
Indies, West, 54.
Infantry, Canadian, 27.
Inganische, 53.
Ingersoll, 47.
International Line, 10, 11, 74, 75.
Iowa, 16.
Ireland, 101.
Irish, the, 61.
Islands, Bay of, 95, 97, 98.
Isle Madame, 54.
— of Dreams, 38.
Isotherm summer, 19.

J.

Jacques Cartier County, 36.
James I, King, 8.
James' Bay, 29.
Jamestown, 8.
Jasper's House, 85.
Java Seas, 82.
Jerseys, 103.
Jesuit, 46.
Joliette, 39.
— County, 36.

K.

Kajoualwang, 32.
Kamloops, 89, 92.
Kamouraska, 32.
— County, 36.
Kananaskis, 85.
Kansas, 15, 16.
Keewatin District, 68.
Kempt Lake, 32.
Kennebecasis, 59.
Kent, Duke of, 35, 65.
Kent County, England, 116.
— New Brunswick, 61, 64.
— Ontario, 47, 49.
Kentville, 56.
Kicking Horse Pass, 85, 88.
King's County, Nova Scotia, 55.
— Prince Edward Island, 66.
Kingsford, the historian, 5.
Kingston, 12, 45, 118.
Knee Lake, 79.
Knights of Nova Scotia, 52.
Kootenay Lake, 90.
— River, 85.

L.

L'Agulhas current, 82.
L'Assomption, 36, 40.
Labrador, 5, 6, 7, 9, 24, 30, 76, 94, 98.
Lachine, 39.
La Collotte, 32.
Ladoga wheat, 121.
Lakes, Canadian, 2, 15.
— of British Columbia, 90.
— of Manitoba, 69.
— of New Brunswick, 60.
— of Nova Scotia, 54.
— of Ontario, 48.
— of Quebec, 32.

Lakes of the North-West, 78.
Lanarkshire, 56.
Lapoile District, 100.
La Prairie County, 36.
Laurentides, 30.
Laurentian Mountains, 30, 31, 96.
Laval, Bishop, 38.
— County, 36.
Leather Pass, 85.
Lethbridge, 80.
Levis, 39.
— County, 36.
Lewis Island, 98.
Liard, F., 19, 82.
Lilloet-Clinton District, 92.
Lilloet Lake, 90.
Lisgar County, 72.
L'Islet County, 36.
Liverpool, 4, 45, 93.
— Nova Scotia, 56, 58.
Logan Mountain, 32.
Logan, Sir W., 5, 31.
London, 12, 56, 68, 116.
— Ontario, 47.
Londonderry, 57.
Longfellow, 52.
Long Island, 19.
— Newfoundland, 98.
— Nova Scotia, 54.
Longueil, 39.
Loon Lake, 60.
Lorne, Marquis of, 44, 48, 56.
Lotbinière County, 36.
Louis XIV, 10.
Louisburg, 51.
Louisiana, 5, 10.
Louisville, 40.
Loyalists, United Empire, 40.
Lumbering, 25.
Lunenburg, 54, 56.
Lyell, Sir Charles, 46, 57, 104.
Lytton, 92.

M.

Macdougall, Lake, 79.
Macfie, Mr., 12.
Mackenzie River, 14, 20, 77, 78, 81, 87, 88, 89.
Mackenzie, Sir J., 19, 86.
MacLeod, Fort, 81.
Macoun, Professor, 19, 104, 105.
Madawaska, 59, 60, 61.

General Index.

Madeira, 68.
Magdalen Island, 29.
Maidstone, Ontario, 47.
Main Factory, 81.
— Gut, 98.
Maine River, East, 29.
Maine, State of, 9, 11, 59, 62, 63, 115.
Maisonneuve, 38.
Maitland, 58.
Maize, 19.
Malacca, Straits of, 82, 83.
Malaga, Lake, 54.
Malicites, 58, 61.
Malta, 24.
Manicouagan, 35.
Manitoba Lake, 16, 43, 70, 90.
— Province, 11, 13, 18, 19, 25, 27, 42, 67, 79.
Manitou, 67.
Manouan, 32.
Manufactures, Canadian, 107.
Maquapit Lake, 60.
Marquette, 72.
Marshall, 46.
Maryland, 9.
Maskinonge County, 36.
Massachusetts, 55.
Massawippi, 40.
Matane River, 32.
Maury, Lieut., 82.
McGill University, 123.
Medicine Hat, 80.
Megantic County, 36.
— Lake, 35, 40.
Melbourne, 3, 12.
Memphramagog, Lake, 35, 40.
Mennonites, 69, 114.
Mercier, Honoré, 40.
Mersey River, Nova Scotia, 53.
Metropolitan of Canada, 61.
Mexico, Gulf of, 10, 24, 55, 77.
— Plain of, 17.
Michigan, Lake, 16.
Micmacs, 58, 61.
Middlesex County, Ontario, 49.
— Ontario, 47.
Milford Haven, 54.
Military Districts, 27.
Milton, 47.
Mines, Basin of, 52, 54.
Mines, Canadian, 22.
Minnedosa, 73.

Minnesota, 17, 18, 33, 70.
Miquelon Island, 97.
Miramichi Bay, 60.
— Lake, 60.
— River, 59.
Miscou Island, 60.
Missisiquoi County, 36.
Mississippi, 10, 26, 77.
Missouri, 16, 77.
Mistassinnie Lake, 32.
Monck County, 49.
Moncton, 62.
Money Orders, 116.
Montana, 15.
Montcalm, 37.
— County, 36.
Montmagny, 39.
— County, 36.
Montmorency, 34.
— County, 36.
Montreal, 5, 12, 13, 16, 20, 30, 33, 39, 45, 48, 68, 80, 117.
— County, 36.
— Island, 30.
Moose Fort, 81.
— Jaw, 80.
— River, 78.
Mozambique current, 82.
Mulagash, Cape, 54.
Municipal Government, 117.
Murchison, Sir R., 31.
Muskoka, Lake, 49.
Musquodoboit Harbour, 54.

N.

Nanaimo, 87, 92, 93.
Nance, Mademoiselle, 39.
Napierville County, 36.
Napoleon I, 10.
Nappan, 104.
Narragansetts, 6.
Nasse River, 87.
Natal, 102.
Native question, 102.
Nebraska, 15, 17.
Nelson County, Vancouver, 92.
— River, 14, 70, 77, 78, 79.
Nepisiguit River, 59, 60.
New Brunswick, 7, 9, 10, 11, 13, 27, 29, 58, 91.
Newcastle, 61.
New England, 8, 9.

New France, 5, 6, 8, 9, 10, 37.
Newfoundland, 4, 5, 6, 7, 9, 10, 24, 26, 28, 63, 83, 93.
New Netherlands, 9.
New South Wales, 12.
New Spain, 10.
New York, 4, 24, 45.
— County, 18.
New Westminster, 85, 91, 93.
New World Island, 98.
Niagara, 33, 46, 47, 109.
— route, 44.
Nicolet, 39.
— County, 36.
Nile Expedition, 27.
Nipigon Lake, 49.
Nipissing Lake, 44, 49.
Noire River, 34.
Norfolk County, Ontario, 49.
Norman County, 94, 97.
North Cape, 16, 24.
North Indian Lake, 79.
North Mountain, 53.
Northumberland County, Ontario, 49.
— New Brunswick, 61, 64.
— Strait, 52, 54, 58, 60, 62.
North-West Territories, 11, 13, 73.
Norumbega, 7.
Norway, 12, 16, 24, 82.
Notre Dame Bay, 95, 97, 98.
— Mountain, 32.
Nottawasaga Bay, 43.
Nouvelle Bretagne, 5.
Nova Scotia, 9, 10, 11, 13, 26, 27, 51, 83.
Nova Zembla, 24.

O.

Oats, Canadian, 19.
Observatory Inlet, 87.
Ohio, 16, 18.
Okanagan, Lake, 90, 92.
Olympia, Mount, 84.
Omineca, 86, 87, 88.
Onega wheat, 121.
Ontario, Lake, 11, 16, 33, 42, 47.
— Province, 11, 13, 18, 27, 42, 49, 53, 78.
Oregon, 10, 20, 88, 93.
— pine, 91.
Orleans, island of, 30.

Oromocto, 59, 60, 61.
Ottawa, city of, 44, 119.
— County, 36.
— River, 4, 12, 29, 30, 32, 33, 41, 42, 45, 81.
Outardes River, 35.
Oxford, 46.
— County, Ontario, 47, 49.
— Street, Ontario, 47.

P.

Pacific, the, 4, 11, 14, 24, 82, 84, 87, 88, 90.
Pacific slope, the, 15, 84.
Pall Mall, Ontario, 47.
Park, a national, 80.
Parliament House, 38.
Parrsborough, 58.
Parsnip River, 86.
Passamaquoddy Bay, 60.
Passes, mountain, 85.
Patent, King James's, 8.
Peace River, 19, 77, 81, 86.
Pedee River, Great, 9.
Peel, 49.
Pelly, Fort, 81.
— Lake, 79.
— River, 77.
Pemberton, 92.
Pennsylvania, 18, 80.
Perth County, Ontario, 47, 49.
Petchora barley, 121.
Peterborough, 49.
Petit Nord Bay, 97.
Petite Nation River, 34.
Petroleum, 108.
Philippines, the, 82.
Philpot, Dr., 25.
Piccadilly, Ontario, 47.
Pictou, 53, 54, 56.
Pilley's Island, 98.
Pine, Cape, 96.
Pines, Canadian, 21.
Pistolet Bay, 97.
Placentia Bay, 97.
— County, 100.
Plains of Abraham, 37.
Point à Beaudet, 29.
— Fortune, 29.
— Lake, 78.
— Levis, 38.
— St. Regis, 29.

Pointe des Monts, 14, 33.
Police, Mounted, 80.
Pontiac County, 36.
Poplar Point, 72.
Port Hawkesbury, 58.
— Hood, 57.
— Medway, 58.
— Nelson, 4.
— Royal, 8, 52.
Portage la Prairie, 72, 73.
Port-à-Port Bay, 96, 97.
Portland, 12, 62, 84.
Portneuf County, 36.
Portneuf River, 35.
Ports of Ontario, 51.
— of Quebec, 41.
Portugal Cove, 7, 96.
Portuguese, 6.
Potomac, 8.
Prairie, the, 25, 26.
Primavista, Cape, 96.
Prince Albert, 70, 81.
— County, 66.
Prince Edward Island, 10, 11, 13, 27, 29, 64, 65, 98.
Protestants, 57, 61, 101, 117, 118.
Provencher County, 72.
Provinces, the, 27.
Pugwash Harbour, 54, 58.

Q.

Qu'Appelle Valley, 74, 80, 81.
Quebec, Province of, 11, 13, 28, 35, 63, 78.
— city, 12, 13, 16, 27, 31, 34, 37.
— County, 36.
Queen Charlotte Island, 28, 83, 84, 85, 87.
Queen's County, New Brunswick, 61.
— Prince Edward Island, 66.
Quesnelle Lake, 90.
Quesnelly, 92.

R.

Race, Cape, 96, 101.
Railway, Grand Trunk, 45.
— Canadian Pacific, 76, 79, 80, 81, 88, 89, 110.
— Intercolonial, 62, 63.
— North-West, 75, 76.
Railways of New Brunswick, 62.

Rainfall of Canada, 17.
Rainy Lake, 49.
— River, 11.
Raleigh, Sir W., 7.
Ray, Cape, 94, 97.
Red Deer River, 76.
Red Indian Lake, 96.
Red River, 14, 20, 67, 69, 70, 73, 78.
Regent Street, Ontario, 47.
Regina, 76, 79, 80.
Reindeer Lake, 16.
Religious sects, 118.
Reserves, Canadian, 27.
Resolution, Fort, 81.
Restigouche, 29, 58, 59, 60.
— County, 61, 64.
Revenue, Dominion, 109.
Rey, Lake, 79.
Richelieu County, 36, 37.
— River, 4, 30, 35.
Richibucto River, 59, 61.
Richmond, 37.
— Bay, 65.
Rideau Canal, 45.
Riding Mountain, 68.
Riel, Louis, 80.
Rifles, Canadian, 27.
Rimouski, 40.
— County, 36.
Riviere du Loup, 63.
Roanoke, 8.
Roberval, Lord of, 7, 8.
Rochester, Ontario, 47.
Rockies, the, 68, 70, 72, 74, 75, 76, 77, 79, 80, 81, 82, 84, 85, 88.
Rockwood, 72.
Rosario Strait, 84.
Rossignol Lake, 54, 58.
Rouge River, 34.
Rousseau Lake, 49.
Rouville County, 36.
Rupert River, 32, 78.
Rupert's Factory, 81.
— Land, 67.
Russia, 11.
Rye, 19, 121.

S.

Saanich Peninsula, 92.
Sable, Cape, 54.
— Island, 54.

Sackville, 62.
Sacred Bay, 97.
Saguenay County, 35.
— River, 4, 30, 32, 34, 40.
St. Agathe, Manitoba, 72.
St. Andrew's, Manitoba, 72.
St. Andrew's, 61.
St. Anne's Bay, 53.
St. Anne's Mountain, 31, 32.
St. Anne, Manitoba, 72.
St. Barbe Bay, 97.
St. Boniface, Manitoba, 72.
St. Catherine's, 47.
St. Charles, Manitoba, 72.
St. Charles River, 37.
St. Clair, 33, 49.
St. Clement's, Manitoba, 72.
St. Croix, 8, 9, 59.
St. Francis River, 4, 35.
St. François Xavier, 72.
St. George's Bay, 63, 96, 97.
St. George, Cape, 54.
St. Hyacinthe, 36, 37.
St. James, Manitoba, 72.
St. Jérôme, 39.
St. John, city of, 12.
St. John's County, New Brunswick, 61.
St. John's Harbour, 60, 101.
St. John's Island, 10, 65.
St. John's Lake, 30, 31, 32, 34.
St. John, New Brunswick, 61, 62, 63, 98.
St. John's County, Newfoundland, 100.
St. John's, Newfoundland, 95, 96, 98.
St. John's County, Quebec, 36, 37.
St. John's River, 10, 59, 60, 61.
St. John's River, Quebec, 35.
St. Lawrence, Cape, 54.
St. Lawrence, Gulf of, 7, 24, 52, 94.
St. Lawrence River, 1, 4, 5, 7, 14, 28, 32, 33, 34, 35, 37, 45, 49, 52, 77, 90.
St. Lawrence Valley, 10, 26.
St. Louis, 33.
St. Mary's, 33.
St. Mary's Bay, 54, 97.
St. Mary's La Have, 53.
St. Margaret Bay, 97.

St. Maurice County, 36.
St. Maurice River, 4, 32, 35, 37, 40.
St. Narbert, Manitoba, 72.
St. Paul's Island, 54.
St. Paul's, Manitoba, 72.
St. Paul's, Ontario, 47.
St. Peter's Lake, 30, 32, 35.
St. Pierre Island, 97.
St. Stephen, 62.
St. Vital, Manitoba, 72.
Salmon River, 60.
Saltcoats, 75, 80.
Salt Spring Island, 92.
Sambro Head Cape, 54.
Sandhurst, a Canadian, 118.
Sandwich, Ontario, 47.
Sandy Lake, 79.
San Francisco, 84, 93.
San Juan Archipelago, 84.
Santee River, 9.
Saskatchewan District, 68, 76, 78, 81.
— River, Little, 70, 76.
Saugeen River, 47.
Sault Ste. Marie, 43, 48.
Savannah River, 9.
Saxonka wheat, 121.
Scarborough, Ontario, 47.
Scatari Island, 54.
Schools, Canadian, 122.
Scots, the, 3, 47, 69, 75.
Selkirk Colonists, 4, 67.
— Earl, 66.
— Mountains, 76, 86, 88, 89.
— town of, 72.
Sequin Lake, 79.
Setting Lake, 79.
Severn Factory, 81.
— Lake, 79.
— River, 79.
Shakespeare, Ontario, 47.
Shasta, 87.
Shecoubish Lake, 32.
Shediac, 62.
— Bay, 60, 66.
Sheet Harbour, 54.
Shelburne, 56.
Shepody Bay, 59, 60.
Sherbrooke County, 36, 37, 63.
Ship Harbour Lake, 54.
Shippegan Island, 60.

Shipping, Canadian, 107.
— Quebec, 40.
Shoal Lake, 73.
Shorthorns, 103.
Shubenacadie River, 53, 58.
Shuswap Lake, 89, 90.
Sicamous Narrows, 89.
Sierra Nevada, 85.
Simcoe County, 49.
— Lake, 49.
Similkameen, 91.
Simpson, Fort, 19, 20, 81, 87.
Skeena River, 84, 86, 87, 88.
Slave Lake, the Great, 16, 77, 78, 79.
— River, 77.
Soils, Canadian, 25.
Sorel, 39.
Soudan, the, 27.
Soulanges County, 36.
Souris River, 66, 70.
South Indian Lake, 79.
South Mountain, Nova Scotia, 53.
Spain, 6.
Spear, Cape, 94, 96.
Spitzbergen, 24.
Split, Cape, 54.
Springfield, 72.
Stanstead County, 36.
Steppes, Prairie, 4.
Stikeen River, 84.
Stony Mountain, 15.
Stratford, Ontario, 47.
Stuart Lake, 90.
Suez Canal, 43.
Suffolk, England, 116, 120.
Sumatra, 83.
Summersberry, 81.
Summerside, 66.
Sunbury County, 61.
Superior, Lake, 4, 11, 14, 16, 33, 48, 71.
Susquehanna, 9.
Sweden, 12.
Sydney, 3, 12.
— Cape Breton, 57.

T.

Tâche, Lake, 79.
Tacla, Lake, 90.
Tadousac, 9, 34.
Tancock Island, 54.

Telegraphs, 101, 116.
Temiscamingue, 20, 29, 32.
Temiscouata County, 36.
— Lake, 59, 60.
Terrebonne, 40.
Texas, 15.
Thames River, Ontario, 47.
Thompson River, 89, 92.
Three Rivers, 33, 37.
— County, 36.
Thunder Bay, 11, 71.
Thutage, Lake, 90.
Tidal action, 109.
Tignish, 65, 66.
Tobique, 59, 60.
Tooke, Vancouver, 92.
Torbay, 54, 58.
Toronto, 12, 13, 17, 20, 24, 45, 47, 117.
Tourmente, Cape, 31.
Townships, Quebec, 40, 41.
— of the North-West, 112.
Trees, Canadian, 25.
Trepassy Bay, 97.
Trinity Bay County, Newfoundland, 100.
Trinity, Cape, 34, 97, 101.
— Gut, 98.
Triton Island, 98.
Trout Lake, 79.
Truro, 56.
Turtle Mountain, 68.
Twelve Mile Creek, 47.
Two Mountains County, 36.
Twillingate, 100.

U.

United States, 10, 11, 12, 15, 16, 20, 29, 43, 54, 58, 83, 84.
Universities, Canadian, 123.
Utrecht, Treaty of, 10, 52.

V.

Valley Field, 39.
Vancouver Island, 4, 16, 56, 83, 84, 86, 87, 91.
Vaudreuil County, 36, 37.
Verchères County, 36.
Vermillion, Fort, 19.
— Pass, 85.
Vermont, 32, 40.
Verrazano, 6.

Viceroy of Canada, 9.
Victoria, Colony of, 12.
— Bridge, 30.
— County, New Brunswick, 61.
— — Nova Scotia, 54.
— — Ontario, 49.
— — Vancouver, 7, 92.
— Falls, 46.
Vimont, 39.
Virginia, 9, 115.
Voltaire, 10.
Voyageurs, Canadian, 27.

W.

Walkerton, 47.
Washademoak, 59.
— Lake, 60.
Washington Lake, 83, 84, 85, 93.
Water-Hen River, 70.
Waterloo County, 49.
Wealth, Canadian, 111.
Welland Canal, 47.
— County, 49.
Wellington County, 49.
Wentworth County, 49.
Westbourne, 72.
Westminster, Ontario, 47.
Westmoreland County, New Brunswick, 61.
Weymouth, 58.
— pine, 91.
Whale River, Great, 78.
Wheat averages, 18.
Whitby, Ontario, 47.
White Bear Bay, 97.
Wigwam River, 85.

William, Fort, 20.
— Creek, 92.
Willis Reach, 98.
Wilson, Sir Daniel, 123.
Windsor, Nova Scotia, 56.
— Ontario, 47.
Winnipeg, city of, 12, 13, 15, 16, 17, 20, 68, 69, 73.
— Lake, 14, 16, 69, 77, 78.
— River, 14, 70, 71.
Winnipegosis Lake, 16, 70.
Wisconsin, 18.
Wolfe, 37.
Wollaston Lake, 78.
Wolseley, Lord, 27.
Woods, Lake of the, 5, 10, 11, 49.
Woodstock, Ontario, 47, 59, 61.

Y.

Yale County, 92.
— Rapids, 91.
Yamaska County, 36.
— River, 35.
Yarmouth, 56.
Yellow Head Pass, 85, 87.
York, Ontario, 45.
— Archipelago, 78.
— County, New Brunswick, 61, 64.
— Factory, 70, 81.
Yorkton, 75.
Yosemite Valley, 87.
Yukon River, 19, 77, 88.

Z.

Zambesi Falls, 46.

THE END.

www.ingramcontent.com/pod-product-compliance
Lightning Source LLC
Chambersburg PA
CBHW020241170426
43202CB00008B/179